CARIBBEAN school atl

T0173241

for Social Studies, Geography and History

Editorial Advisor: Professor Michael Morrissey

HODDER EDUCATION
AN HACHETTE UK COMPANY

These small maps explain the meaning of some of the lines and colours on the atlas maps.

LAND

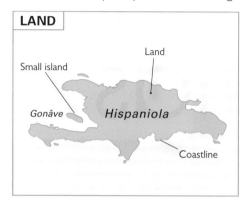

This is how an island is shown on a map. The land is coloured green. The coastline is a blue line.

OCEANS & SEAS

There are 5 large areas of water on Earth called oceans. Smaller areas are called seas, such as the Caribbean Sea.

RIVERS & LAKES

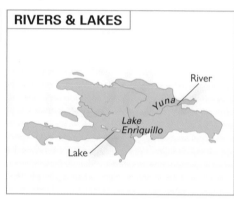

Areas of water on land are shown in a blue tint and are called lakes. Rivers are shown as blue lines.

HEIGHT OF THE LAND

Beside a topographical map is a diagram to show the different colours used for the height of the land.

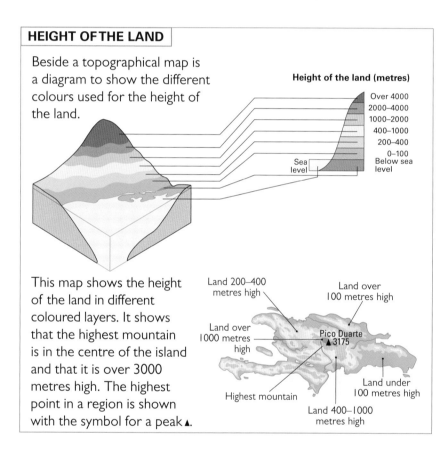

Height of the land (metres)

Over 4000
2000–4000
1000–2000
400–1000
200–400
0–100
Below sea level
Sea level

This map shows the height of the land in different coloured layers. It shows that the highest mountain is in the centre of the island and that it is over 3000 metres high. The highest point in a region is shown with the symbol for a peak ▲.

COUNTRIES, CITIES & TOWNS

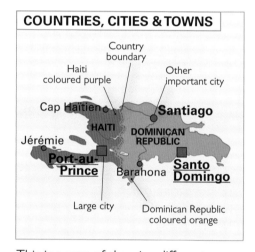

This is a way of showing different information about the island. It shows that the island is divided into two countries. They are separated by a country boundary. There are cities and towns on the island. The two capital cities are underlined. Large cities are shown by a red square. Other important cities are shown by a red circle.

TRANSPORT INFORMATION

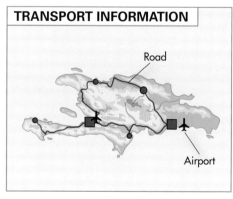

This map shows the most important roads and airports. Transport routes connect the cities and towns. Airports are shown by an aeroplane symbol.

WHERE IS THE ISLAND?

This map gives lines of latitude and longitude. These show where the island is in the world. Latitude and longitude are explained on page 5.

A COMPLETE MAP

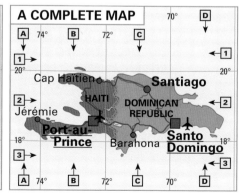

This map is using the country colouring and shows the letter-number codes used in the index. See page 6 on how to use these codes.

Every map is designed for a specific purpose. The general maps on page 2, opposite, show for a place the height of the land and the rivers (topography). They also show the towns and cities, the roads, railways and airports. The boundaries between countries and administrative regions are shown.

Another kind of map is specially designed to show a specific topic or theme. This is a thematic map. This page has five examples of thematic maps in your atlas. Some are taken from the Caribbean Region section of the atlas, some from Caribbean Countries, and others from the Continents & Countries section. Read about the theme shown, look at the complete map on the correct atlas page, and then answer the questions.

MAJOR HURRICANES The theme of this map of the Caribbean is major hurricanes since 2004. Five are shown in the extract, starting in the Atlantic Ocean and moving west. The track taken by each of these hurricanes is shown with a different colour. Now look at the complete map on page 13.
- Can you see some other hurricanes which began in other places?
- How many of these hurricanes passed over The Bahamas?
- Do you know names of hurricanes which are not shown on this map?

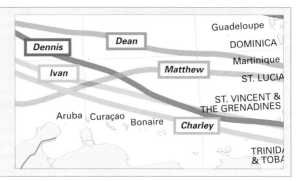

LIFE EXPECTANCY The theme of this map is how many years of life is expected for someone born in 2016. The dark green is for countries where the newly born child can expect to live for longer than average - for more than 80 years. If you live in a country shaded pale yellow, the average person will not reach 70 years of age. Now look at the complete map on page 16.
- What is the life expectancy of a baby born in your country in 2016?
- Which three Caribbean countries have the lowest life expectancy?
- Why do some countries have a lower life expectancy than others?

TOURISM This map uses colours, symbols and numbers to show which parts of a country are important for the tourism industry. The main tourism resort areas are shaded green. Hotels and attractions are shown by symbols. Special places visited by tourists are numbered. Now look at the complete map on page 32.
- Where do most tourists stay in Jamaica?
- Where do cruise ships stop?
- Why do some tourists visit the village where Bob Marley was born?

CLIMATE TYPES Each colour on this map of the world shows a type of climate. Nine climate types are shown on this map but there are many other ways to classify climates. For example, on this map yellow shows a Savanna type of climate which has high temperatures and a long dry season. Look at the complete map on page 78.
- How many regions of the world have a humid tropical climate?
- How many climate types does Australia have?
- What kind of climate does Brussels, in Europe, have?

RESOURCES This map shows the gas and oil resources in the south of Trinidad & Tobago on-shore and off-shore. Oil fields are shown in green. Gas fields are shown in brown. Pipelines link the fields to refineries and tanker terminals. Look at the entire map on page 52.
- What is the name of the gas field nearest to Tobago?
- Where is the oil refinery?
- Which country does the Dragon oil field belong to?

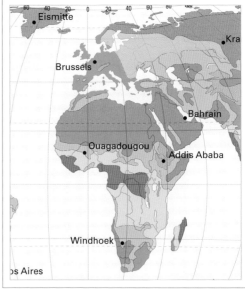

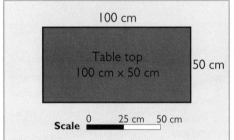

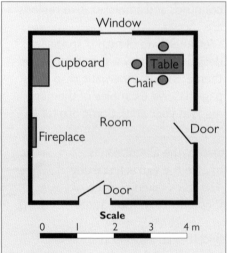

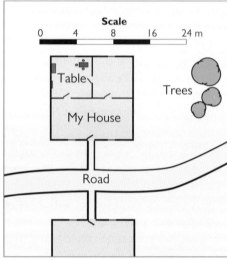

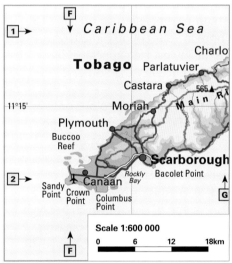

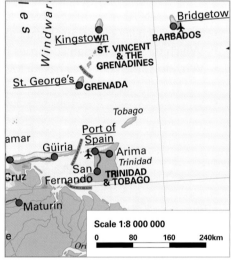

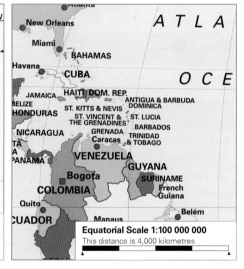

The drawing of the top of a table is looking down on it. It is 100 cm long and 50 cm wide. The drawing measures 4 x 2 cm. It is drawn to scale: 1 cm on the drawing equals 25 cm on the table.

This is a plan of a room looking down from above. 1 cm on the plan equals 1 metre in the room. The same table is shown, but now it is shown at a smaller scale.

This is an even smaller scale plan that shows the table in a room, inside a house. 1 cm on the plan equals 4 metres in the house. We can also call this a large scale map.

This is a map that you will find in this atlas. It is a medium-scale map and shows much more detail than the maps to its right. The map above is a larger-scale map than the maps on the right.

This is also a map from the atlas. It is at a smaller scale than the map on the left. Hence, a large-scale map shows more detail of a small area. A small-scale map shows less detail for a larger area.

This is part of the map on pages 58 and 59 in this atlas. The scale is small enough to show the whole world on two pages. This map shows the relative sizes of countries.

TYPES OF SCALE

In this atlas the scale of the map is shown in three ways.

Written Statement	**Ratio**	**Scale Bar**
This tells you how many kilometres on the earth are represented by one centimetre on the map.	This tells you that one unit on the map represents two million of the same units on the ground	This shows you the scale as a bar with a section of a ruler beneath it.
1cm on the map = 20km on the ground	**Scale 1:2 000 000**	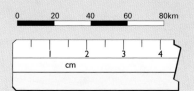

NORTH POINT

This map has a North Point showing the direction of north. It points in the same direction as the lines of longitude.

THE CARDINAL POINTS

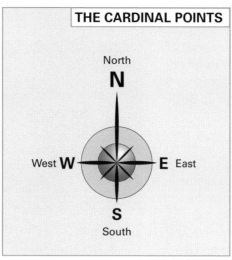

The North Point shows the four main directions: north, east, south and west. These are called the cardinal points.

THE INTERCARDINAL POINTS

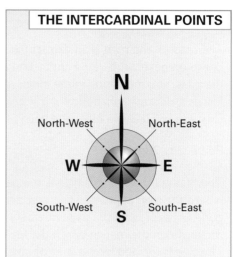

This diagram shows the intercardinal points. For example, between North and East is North-East.

LATITUDE

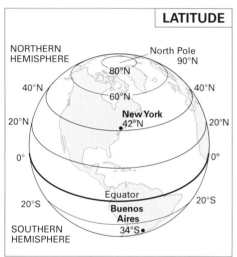

Lines of latitude cross the atlas maps from east to west. The Equator is at 0°. All other lines of latitude are either north or south of the Equator. Line 40°N is almost halfway towards the North Pole. The North Pole is at 90°N. What is the latitude of your country?

LONGITUDE

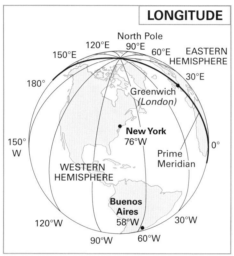

Lines of longitude run from north to south. These lines meet at the North Pole and the South Pole. Longitude 0° passes through Greenwich. This line is also called the Prime Meridian. Lines of longitude are either east of 0° or west of 0°. There are 180 degrees of longitude both east and west of 0°.

USING LATITUDE & LONGITUDE

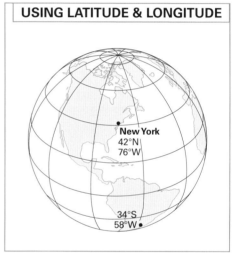

Latitude and longitude lines make a grid that can be printed on a map. You can find a place if you know its latitude and longitude. The degree of latitude is either north or south of the Equator. The longitude number is either east or west of the Greenwich Meridian.

WHICH LINES OF LATITUDE ARE IMPORTANT?

The map on the right shows the lines of latitude which are especially important. Across the middle of the map runs the Equator which is the starting point for measuring all other lines of latitude. There are two lines called the 'Tropics'. North of the Equator, the Tropic of Cancer marks the place where the sun is overhead at midsummer in the northern hemisphere. South of the Equator, the Tropic of Capricorn marks the place where the sun is overhead at midsummer in the southern hemisphere. In the North and South Polar regions, the Arctic Circle and the Antarctic Circle show the limits of the area where the Sun does not rise or set above the horizon at certain times of the year.

COPYRIGHT PHILIP'S

HOW TO FIND A PLACE

The map on the right is an extract from the map of Belize on page 23. If you want to find Belize City in the atlas, you must look in the index (pages 86 to 89). Places are listed alphabetically. If you look up Belize City you will find the following entry:

Belize City **23** B2

The first number in bold type is the page number where the map appears. The letter and number code (which follow the page number) give the grid rectangle on the map in which the feature appears. The grid is formed by the lines of latitude and longitude which are shown in blue. The letter and number codes are shown in yellow boxes around the edge of the maps. Here we can see that Belize City is on page 23 in the rectangle where column B crosses row 2.

If you need to find the latitude and longitude of a place you can work this out from a map. The latitude and longitude numbers are in black at the ends of the blue lines on the map. Latitude and longitude are measured in degrees as explained on page 5. The degree (°) is divided into 60 minutes.
Can you work out the latitude and longitude of Belize City?

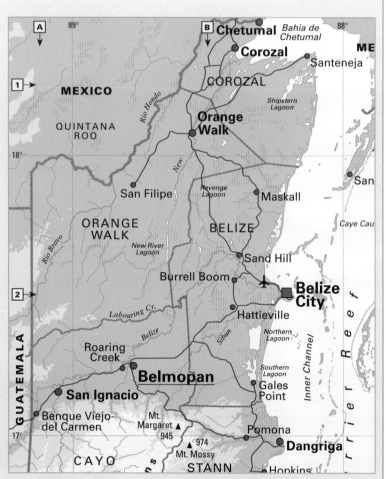

HOW TO MEASURE DISTANCE

The map on the right is a small part of the map of the Caribbean, which is on page 11 in the Caribbean Region section of the atlas.

The scale of the map extract is shown below:

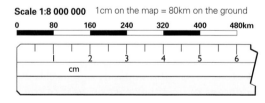

Scale 1:8 000 000 1cm on the map = 80km on the ground

To measure the distance from Port of Spain, Trinidad to Castries, St. Lucia you can use any of the three methods described below.

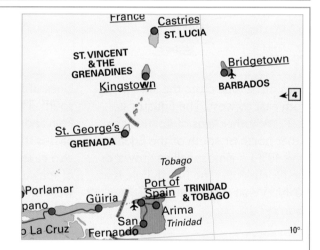

Using the written statement
Using the scale above, you can see that 1 cm on the map represents 80 km on the ground.

Measure the distance on the map between Port of Spain and Castries. You will see that it is about 4.5 cm.

If 1 cm = 80 km

then 4.5 = 360 km (4.5 x 80)

Using the ratio
Using the scale above, you can see that the ratio is 1:8 000 000.

We know that the distance on the map between the cities is 4.5 cm and we know from the ratio that 1 cm on the map = 8 000 000 cm on the ground. We multiply the map distance by the ratio.

= 4.5 x 8 000 000 cm
= 36 000 000 cm
= 360 000 m
= 360 km

Using the scale bar
We know that the distance on the map between the cities is 4.5 cm.

Using the scale bar, measure 4.5 cm along this (or use a ruler as a guide) and read off the distance.

• Using these three methods, now work out the distance between Port of Spain and Bridgetown, Barbados on the map above. Your teacher could tell you if your answer is correct.

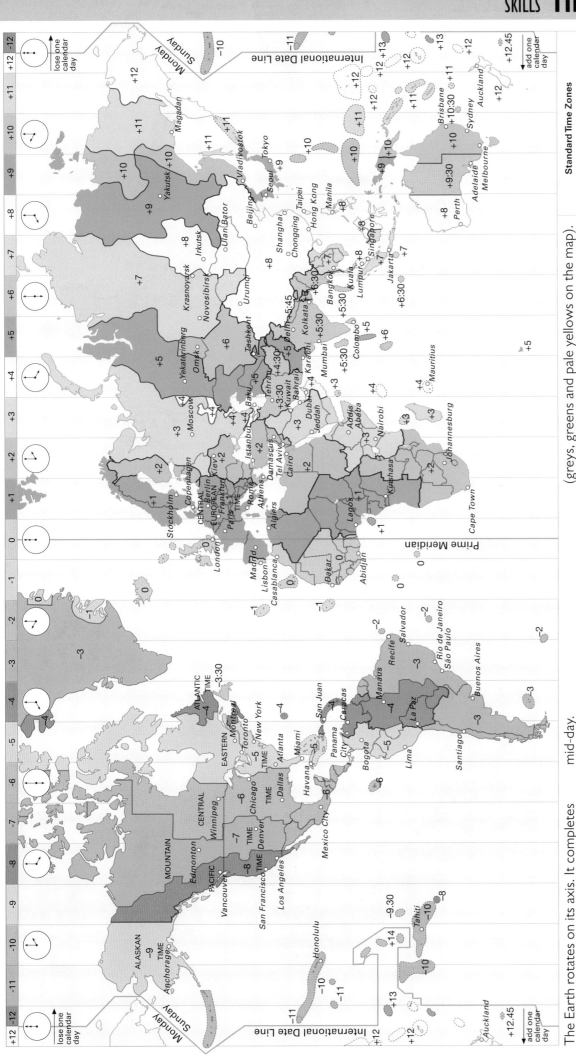

The Earth rotates on its axis. It completes one rotation in 24 hours. Because it rotates through 360 degrees (360°), it therefore rotates though 15 degrees (15°) every hour. Therefore, the earth has been divided into segments of 15 degrees, as the map shows. Each segment is a Time Zone. For each time zone, the sun is at its highest point at around mid-day.

The zero-degree line of longitude (0°) that runs from the north pole to the south pole through Greenwich in London is the standard agreed by all countries. The zone of 15 degrees (15°) either side of the Greenwich meridian has Greenwich Mean Time or GMT. To the east, the zones are ahead of GMT

(greys, greens and pale yellows on the map). To the west, the zones are behind GMT (red and orange tints).

The 180-degree (180°) line of longitude is the International Date Line. The date is different on the each side of this line.

- How many hours behind GMT is your home?

Standard Time Zones

Zones using Greenwich Mean Time
Zones behind Greenwich Mean Time
Zones ahead of Greenwich Mean Time
Hours slow or fast of Greenwich Mean Time
10 Half-hour zones
International boundaries
Time zone boundaries
International Date Line

COPYRIGHT PHILIP'S

GULF OF MEXICO

U.S.A.
The
West Palm Beach
Fort Lauderdale
Everglades
Freeport
Grand
Bahama
Abaco I.

BAHAMAS

Miami

Bimini Is.

Dry Tortugas
(U.S.A.)
Key West
Florida Keys

Florida Strait

Eleuthera I.
Nassau

Great Bahama Bank

Cat I.

Great
Exuma I.
Long I.

Ac

Cay Sal
Bank

Havana
Matanzas

Pinar del Río
Batabanó
Santa Clara
Morón

CUBA

Cienfuegos

I. de la
Juventud

Camagüey

Victoria de
las Tunas

Holguín

Bar

Guantánamo

Santiago
de Cuba

1972

GUANTÁNAMO
BAY
(U.S.A.)

Win

Yucatan Str.
C. San Antonio

Progreso
Tizimín
Cancún

C. Catoche

Mérida

Cozumel
Isla
Cozumel

MEXICO

Cayman Islands
(U.K.)
George Town

Cayman Trench

Navassa I.
(U.S.A.)

Montego Bay

Savanna-la-Mar
JAMAICA
Mandeville
Kingston

Les Cay

Chetumal

Ambergris Cay

Belize
City
Turneffe Is.

Is. Santanilla
(Honduras)

BELIZE

Belmopan

Gulf of Honduras
Is. de
la Bahía
Roatán

Pedro Cays
(Jamaica)

Morant
Cays
(Jamaica)

Puerto
Barrios

La Ceiba

San Pedro Sula

HONDURAS

Central

Bajo Nuevo
(Colombia)

C. Gracias a Dios

Coco (Segovia)

C A R I B B

Cayos Miskitos
(Nicaragua)

Tegucigalpa

EL
SALVADOR
San Miguel

A m e r i c a

Matagalpa

Cayos Roncador
(Colombia)

I. de Providencia
(Colombia)

See cross-section
at bottom of page 9

G. de Fonseca
Chinandega

NICARAGUA

León

Managua

I. de San Andrés
(Colombia)

Santa
Marta

Lake
Nicaragua

Mosquito Coast

Is. del
Maiz
(Nicaragua)

Barranquilla

Sole

COSTA

San José
Limón

Cartagena

Puntarenas
RICA

*Isthmus of
Panama*

G. de Nicoya

3837

Volcan Barú

3374

Panama
Canal
Colón

G. de los
Mosquitos

Gulf of Darién

Sincelejo

PA
N
A
M
A

Panamá

Monteria

Magangué

Mo

PACIFIC
OCEAN

David

Arch.
de las
Perlas

Gulf of
Panama

Yaviza

COLOMBIA

Pen. de
Azuero

I. de Coiba

■ Over 1,000,000 inhabitants
● Under 1,000,000 inhabitants
Panamá Capital cities underlined

—— Roads
✈ Main airports
✈ International boundaries

⊥ Swamps and marshes
Canals

Height of the land (metres)

Over 4000
2000–4000
1000–2000
400–1000
200–400
Sea level 0–200
Below sea
level

Scale 1:8 000 000 1cm on the map = 80km on the ground
0 80 160 240 320 400 480km

Turks & Caicos
(U.K.)
Cockburn
Town

A T L A N T I C   O C E A N

Tropic of Cancer

Mayaguana I.

reat
agua I.

Cap-
Haïtien
Gonaïves
Santiago de
los Caballeros
San Francisco
de Macorís
ITI
au-Prince
Pico Duarte
3175▲
DOMINICAN
REP.
Barahona
Santo
Domingo
La Romana
ispaniola
ntilles

Milwaukee
Deep
▼
9200

Puerto Rico
Trench

San
Juan
Charlotte Amalie
British
Virgin Is.
Anguilla
(U.K.)
St.-Martin(Fr.)
Ponce
PUERTO
RICO
(U.S.A.)
Mona Passage
Isla
Mona
St. Maarten
(Neth.)
St.-Barthélemy (Fr.)
U.S.
Virgin Is.
St. Croix
Saba
(Neth.)
ST. KITTS
& NEVIS
ANTIGUA
& BARBUDA
Basseterre
St. John's
Montserrat
(U.K.)
Guadeloupe
(Fr.)
Pointe-à-Pitre
Basse-Terre
DOMINICA
Roseau
I. de Aves
(Ven.)
Martinique
(Fr.)
Fort-de-
France
Castries
ST. LUCIA
Bridgetown
Kingstown
BARBADOS
ST. VINCENT
& THE
GRENADINES
St. George's
GRENADA

A N   S E A

Leeward Islands

Lesser Antilles

Windward Islands

Pta. Gallinas
Oranjestad
Aruba
(Neth.)
Curaçao
(Neth.)
Bonaire (Neth.)
Willemstad
I. Orchila
(Ven.)
Is. Los Roques
(Ven.)
I. Blanquilla
(Ven.)
Tobago
Pen. de la
Guajira
Pen. de
Paraguaná
Punto Fijo
Nevada de
Marta
acha
Gulf of
Venezuela
Coro
Puerto
Cabello
Caracas
I. de Margarita
(Ven.)
I. La Tortuga
(Ven.)
Porlamar
Güiria
Port of
Spain
Arima
Trinidad
Maracaibo
Cabimas
San Felipe
Maracay
Cumaná
Carúpano
San
Fernando
TRINIDAD
& TOBAGO
dupar
Barquisimeto
Valencia
Puerto La Cruz
Barcelona
Maturín
Lake
Maracaibo
South
Valera
America
Delta of the
Orinoco
El Tigre
Orinoco
Boca Grande
Cord. de Mérida
Barinas
VENEZUELA
Ciudad
Bolívar
Ciudad Guayana
icuta
West from Greenwich
San Fernando
de Apure
Orinoco

GUYANA map (upper right):

A T L A N T I C
O C E A N
VENEZUELA
Orinoco
Cuyuni
Georgetown
Bartica
New Amsterdam
Linden
Corriverton
G U Y A N A
Pakaraima
Kaieteur
Falls
2772▲ Mt.
Roraima
Mts.
Essequibo
Corentyne
SURINAME
BRAZIL
Tacutu
Kanuku Mts.
Boa
Vista
G u i a n a
H i g h l a n d s
GUYANA
on same scale

CROSS SECTION (bottom):

SS SECTION
ros
nd
Great
Bahama
Bank
Cayman Trench
(Deepest point 7680m)
Cuba
Jamaica
Caribbean
Sea
Colombian
Coast
2000m
Sea level
2000m
4000m
6000m
8000m
Ⓑ

GUYANA inset (bottom right):

Cuyuni
Georgetown
Bartica
GUYANA
Linden
Pakaraima
Kaieteur
Falls
2772▲ Mt.
Roraima
Mts.

COPYRIGHT PHILIP'S

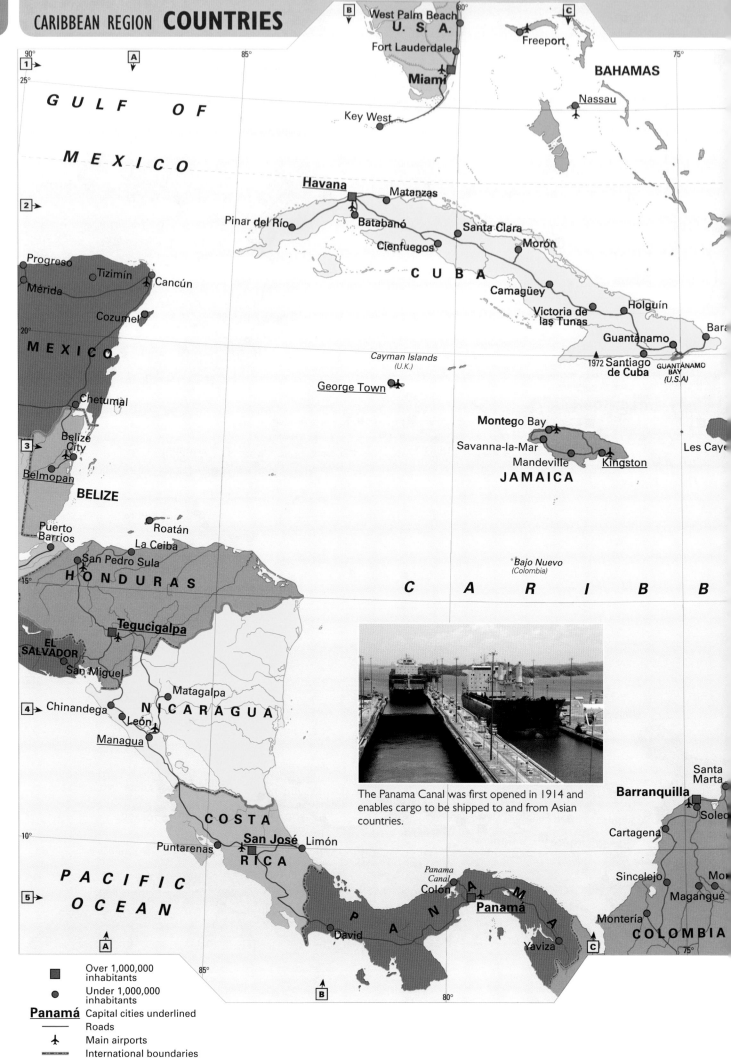

GULF OF

MEXICO

BAHAMAS

West Palm Beach
U.S.A.
Fort Lauderdale
Miami
Key West
Freeport

Nassau

Havana
Matanzas
Pinar del Río
Batabanó
Santa Clara
Cienfuegos
Morón
CUBA
Camagüey
Victoria de
las Tunas
Holguín
Bara
Guantánamo
1972 Santiago GUANTÁNAMO
de Cuba BAY
(U.S.A)

Cayman Islands
(U.K.)
George Town

Montego Bay
Savanna-la-Mar
Mandeville
Kingston
Les Caye
JAMAICA

Progreso
Tizimín
Cancún
Mérida
Cozumel
MEXICO
Chetumal
Belize
City
Belmopan
BELIZE

Puerto
Barrios
Roatán
La Ceiba
San Pedro Sula
HONDURAS

Bajo Nuevo
(Colombia)

C   A   R   I   B   B

Tegucigalpa
EL SALVADOR
San Miguel

Matagalpa
Chinandega
NICARAGUA
León
Managua

The Panama Canal was first opened in 1914 and
enables cargo to be shipped to and from Asian
countries.

Santa
Marta
Barranquilla
Soled
Cartagena

COSTA

San José Limón
Puntarenas
RICA

PACIFIC

OCEAN

Panama
Canal
Colón
P   A   N   A   M   A
Panamá

David

Yaviza

Sincelejo
Mon
Magangué
Montería
COLOMBIA

Legend:

- ■ Over 1,000,000 inhabitants
- ● Under 1,000,000 inhabitants
- **Panamá** Capital cities underlined
- —— Roads
- ✈ Main airports
- ⊢⊣ International boundaries

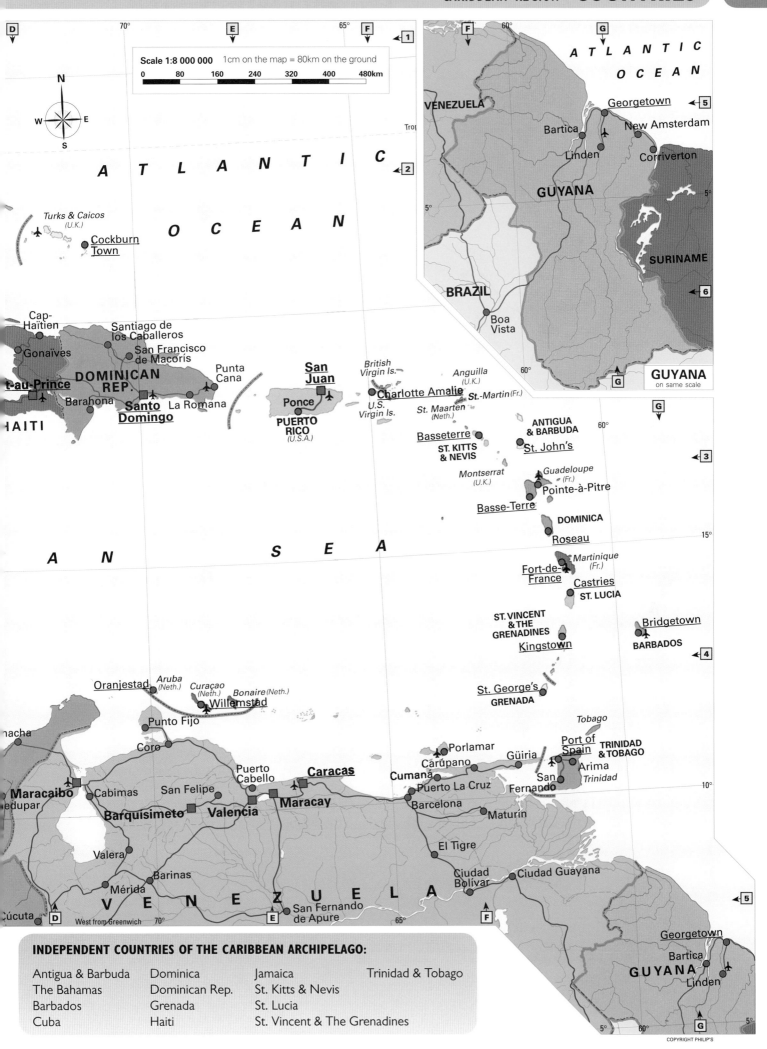

Scale 1:8 000 000   1cm on the map = 80km on the ground

0   80   160   240   320   400   480km

ATLANTIC OCEAN

ATLANTIC OCEAN

VENEZUELA

Georgetown
Bartica
Linden
New Amsterdam
Corriverton

GUYANA

SURINAME

BRAZIL

Boa Vista

GUYANA on same scale

Turks & Caicos (U.K.)
Cockburn Town

Cap-Haïtien
Santiago de los Caballeros
Gonaïves
San Francisco de Macorís
Punta Cana
DOMINICAN REP.
t-au-Prince
Barahona
Santo Domingo
La Romana
HAITI

San Juan
Ponce
PUERTO RICO (U.S.A.)

British Virgin Is.
Charlotte Amalie
U.S. Virgin Is.
St.-Martin (Fr.)
St. Maarten (Neth.)

Anguilla (U.K.)

ANTIGUA & BARBUDA
St. John's

Basseterre
ST. KITTS & NEVIS

Montserrat (U.K.)

Guadeloupe (Fr.)
Pointe-à-Pitre
Basse-Terre

DOMINICA
Roseau

Martinique (Fr.)
Fort-de-France
Castries
ST. LUCIA

ST. VINCENT & THE GRENADINES
Kingstown

Bridgetown
BARBADOS

St. George's
GRENADA

Tobago

Oranjestad
Aruba (Neth.)
Curaçao (Neth.)
Bonaire (Neth.)
Willemstad
Punto Fijo
Coro

Porlamar
Carúpano
Güiria
Port of Spain
TRINIDAD & TOBAGO
Cumaná
Arima
Trinidad
San Fernando

acha
Maracaibo
edupar
Cabimas
San Felipe
Puerto Cabello
Caracas
Maracay
Barquisimeto
Valencia
Puerto La Cruz
Barcelona
Maturín

Valera
El Tigre
Barinas
Mérida
Ciudad Bolívar
Ciudad Guayana
Cúcuta
West from Greenwich
VENEZUELA
San Fernando de Apure

Georgetown
Bartica
GUYANA
Linden

INDEPENDENT COUNTRIES OF THE CARIBBEAN ARCHIPELAGO:

Antigua & Barbuda	Dominica	Jamaica	Trinidad & Tobago
The Bahamas	Dominican Rep.	St. Kitts & Nevis	
Barbados	Grenada	St. Lucia	
Cuba	Haiti	St. Vincent & The Grenadines	

CARIBBEAN PLATE

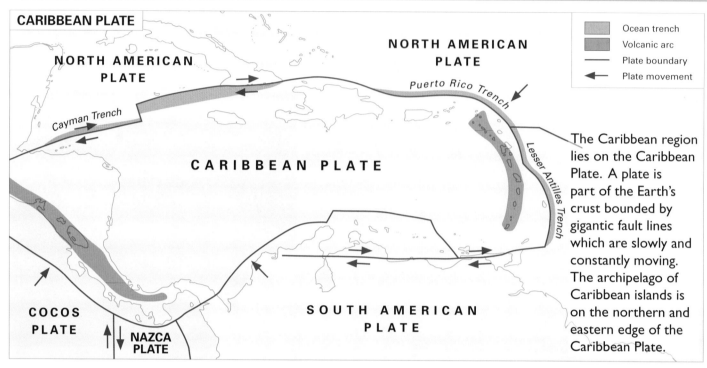

	Ocean trench
	Volcanic arc
	Plate boundary
←	Plate movement

The Caribbean region lies on the Caribbean Plate. A plate is part of the Earth's crust bounded by gigantic fault lines which are slowly and constantly moving. The archipelago of Caribbean islands is on the northern and eastern edge of the Caribbean Plate.

EARTHQUAKES, VOLCANOES & TSUNAMIS

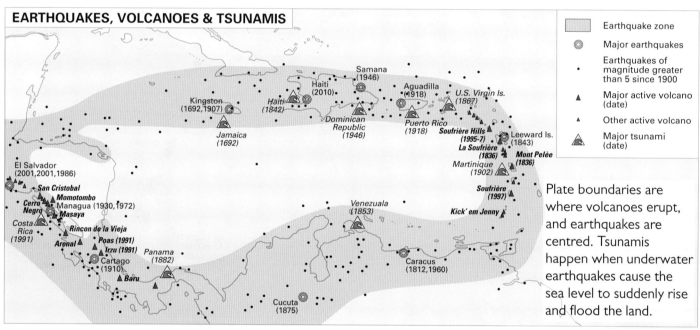

	Earthquake zone
◎	Major earthquakes
•	Earthquakes of magnitude greater than 5 since 1900
▲	Major active volcano (date)
▴	Other active volcano
🜂	Major tsunami (date)

Plate boundaries are where volcanoes erupt, and earthquakes are centred. Tsunamis happen when underwater earthquakes cause the sea level to suddenly rise and flood the land.

PLATE TECTONICS IN THE CARIBBEAN

This cross section gives you an idea of what is happening below the surface. The line of section is along 15 degrees North (15°N). Look at this line of latitude on the Caribbean map on pages 8 and 9. The line crosses Guatemala and Honduras in the west, passes through the Caribbean Sea, then a little south of Dominica and on to the Atlantic Ocean to the east. The diagram shows the relationship of plate boundaries to volcanic eruptions and earthquakes.

• Is your country affected?

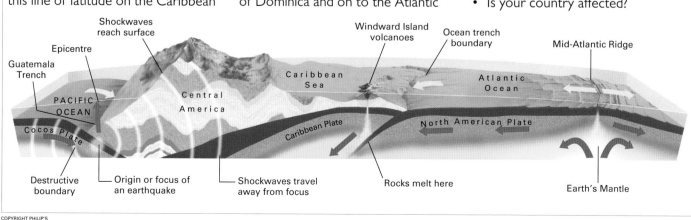

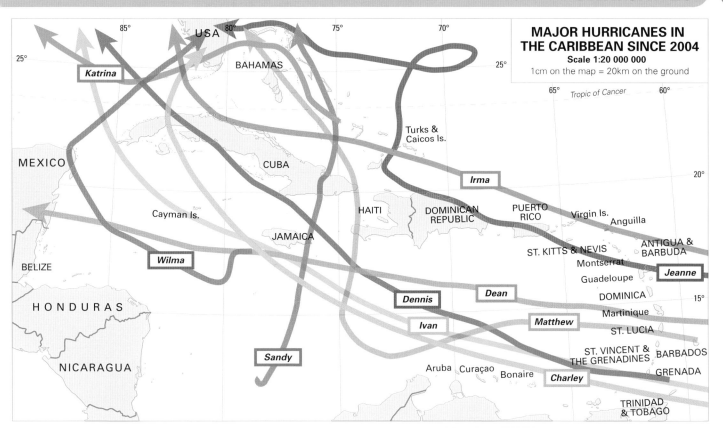

MAJOR HURRICANES IN THE CARIBBEAN SINCE 2004
Scale 1:20 000 000
1cm on the map = 20km on the ground

Name	Places badly affected	Period	Strength	Deaths
Charley	Barbados, Cayman Is., Jamaica, Cuba	9–14 Aug. 2004	4	10
Ivan	Tobago, Grenada, Jamaica, Cayman Is., Cuba	20–24 Sept. 2004	5	92
Jeanne	Guadeloupe, Puerto Rico, Dominican Republic, Haiti	13–28 Sept. 2004	3	3000+
Dennis	Cuba	4–18 Jul. 2005	4	88
Katrina	Bahamas	28–29 Aug. 2005	5	1,836
Wilma	Haiti, Jamaica	19 Oct. 2005	5	87
Dean	Lesser Antilles, Jamaica	18–21 Aug. 2007	5	45
Sandy	Jamaica, Cuba, Bahamas	22–29 Oct. 2012	3	286
Matthew	Haiti, Cuba, Dominican Republic	1 Oct. 2016	5	603
Irma	Barbuda, St Martin, Anguilla, Virgin Is., Turks & Caicos,	30 Aug.–16 Sept. 2017	5	102

SAFFIR-SIMPSON HURRICANE SCALE

Category 1 – Weak hurricane
Winds 119-153 km/hour. Some damage and power cuts.

Category 2 – Moderate hurricane
Winds 154-177 km/hour. Extensive damage and power cuts. Many trees uprooted or snapped.

Category 3 – Strong hurricane
Winds 178-208 km/hour. Well-built homes suffer major damage. Trees uprooted. Flooding inland. Total power loss.

Category 4 – Very strong hurricane
Winds 209-251 km/hour. Catastrophic damage. Severe damage to well-built homes, trees blown over.

Category 5 – Devastating hurricane
Winds over 252 km/hour. Many buildings destroyed, major roads cut off. Damage by storm surge.

CROSS-SECTION THROUGH A HURRICANE

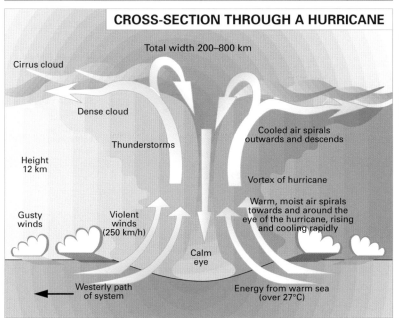

Hurricane Irma (above), with winds of 295km/hour, was the most powerful in over ten years when it made landfall on Barbuda in September 2017. It caused catastrophic damage in St. Barthélemy, St. Martin, Anguilla and the Virgin Islands.

ERA OF EUROPEAN EXPLORATION

The Ciboneys, Arawaks and Caribs (Tainos and Kalinagos) settled the Caribbean long before the arrival of Europeans. It is estimated that at the time of Columbus' voyages there were 200,000 Amerindian peoples in the Caribbean region. Most were killed by the diseases brought by the Europeans.

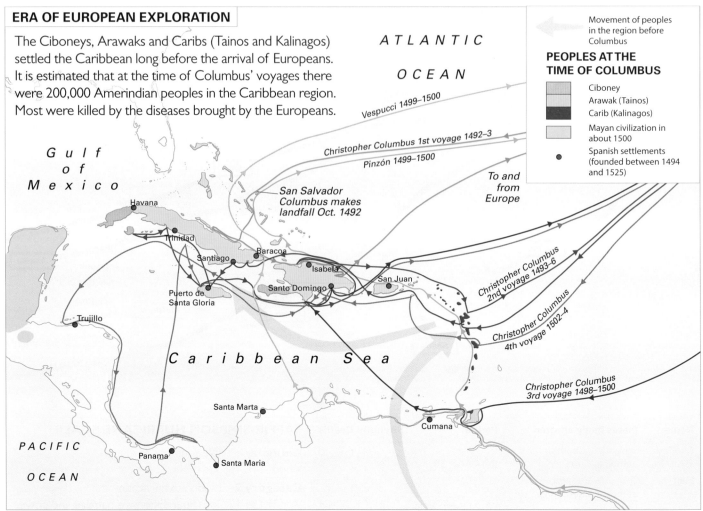

PEOPLES AT THE TIME OF COLUMBUS

- Movement of peoples in the region before Columbus
- Ciboney
- Arawak (Tainos)
- Carib (Kalinagos)
- Mayan civilization in about 1500
- Spanish settlements (founded between 1494 and 1525)

AGE OF SPANISH CONQUEST OF AMERICAN CIVILISATIONS

The Spaniards were mainly interested in the gold and silver of South America. They also colonised islands of the Greater Antilles. Other European countries began to challenge the Spaniards in the Caribbean and to occupy islands in the Lesser Antilles. They cultivated tobacco, indigo, sugar and other crops for export.

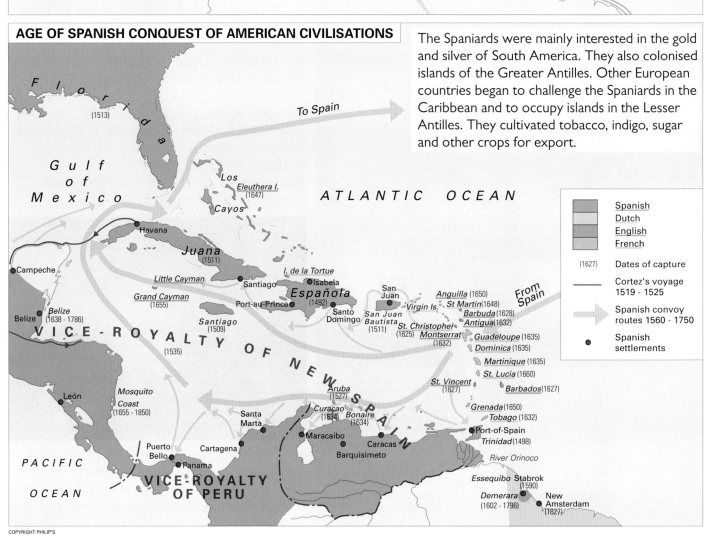

- Spanish
- Dutch
- English
- French
- (1627) Dates of capture
- Cortez's voyage 1519 - 1525
- Spanish convoy routes 1560 - 1750
- Spanish settlements

COLONIES

(1638) Dates of acquisition/independence

Spanish
Dutch
British
French
U.S.

Slave routes and indentured labour routes

Rebellions

● Major towns and naval bases

Today's international boundaries

EUROPEAN SLAVE TRADE & AFRICAN REBELLION

European demand for sugar grew. Over four million Africans were enslaved and shipped to the West Indies. In all, over ten million Africans were taken across the Atlantic in slave ships.

• When was slavery prohibited in your country?

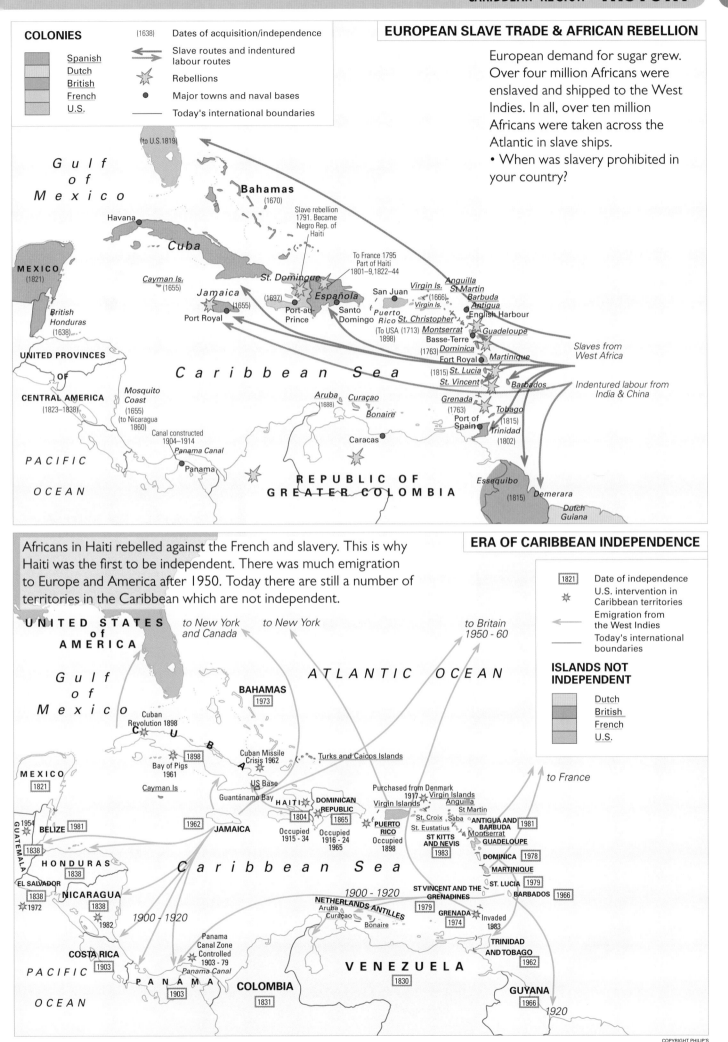

Gulf of Mexico

(to U.S.1819)

Bahamas (1670)

Havana ●

Cuba

Slave rebellion 1791. Became Negro Rep. of Haiti

To France 1795 Part of Haiti 1801–9, 1822–44

MEXICO (1821)

Cayman Is. (1655)

Jamaica (1655)

Port Royal ●

(1697) *Española*

St. Domingue

Port-au-Prince

Santo Domingo

San Juan

Puerto Rico (To USA 1898)

Virgin Is. (1666) *Virgin Is.*

St. Christopher (1713)

Montserrat

Anguilla
St Martin
Barbuda
Antigua English Harbour

British Honduras (1638)

UNITED PROVINCES OF CENTRAL AMERICA (1823–1838)

Mosquito Coast (1655) (to Nicaragua 1860)

Caribbean Sea

Aruba (1688) *Curaçao*

Bonaire

Basse-Terre (1763) *Dominica*

Fort Royal ● *Martinique*

(1815) *St. Lucia*
St. Vincent

Guadeloupe

Barbados

Slaves from West Africa

Indentured labour from India & China

Canal constructed 1904–1914

Panama Canal

Panama ●

PACIFIC OCEAN

Caracas ●

Grenada (1763)

Port of Spain ●

Tobago (1815)

Trinidad (1802)

REPUBLIC OF GREATER COLOMBIA

Essequibo (1815)

Demerara

Dutch Guiana

Africans in Haiti rebelled against the French and slavery. This is why Haiti was the first to be independent. There was much emigration to Europe and America after 1950. Today there are still a number of territories in the Caribbean which are not independent.

ERA OF CARIBBEAN INDEPENDENCE

1821 Date of independence

✳ U.S. intervention in Caribbean territories

← Emigration from the West Indies

Today's international boundaries

ISLANDS NOT INDEPENDENT

Dutch
British
French
U.S.

UNITED STATES of AMERICA

Gulf of Mexico

to New York and Canada

to New York

ATLANTIC OCEAN

to Britain 1950 - 60

BAHAMAS 1973

Cuban Revolution 1898

C U B A 1898

Bay of Pigs 1961

Cuban Missile Crisis 1962

Turks and Caicos Islands

Cayman Is

US Base Guantánamo Bay

MEXICO 1821

HAITI 1804

DOMINICAN REPUBLIC 1865

Occupied 1915 - 34

Occupied 1916 - 24 1965

JAMAICA 1962

PUERTO RICO Occupied 1898

Virgin Islands

Purchased from Denmark 1917 Virgin Islands

St. Croix Saba
St. Eustatius

Anguilla
St Martin

ANTIGUA AND BARBUDA 1981

Montserrat

ST KITTS AND NEVIS 1983

GUADELOUPE

DOMINICA 1978

MARTINIQUE

to France

GUATEMALA 1954

BELIZE 1981

1838

HONDURAS 1838

EL SALVADOR 1838

NICARAGUA 1838 1972 1982

1900 - 1920

Panama Canal Zone Controlled 1903 - 79

Panama Canal

COSTA RICA 1903

PANAMA 1903

PACIFIC OCEAN

COLOMBIA 1831

Caribbean Sea

1900 - 1920

NETHERLANDS ANTILLES
Aruba Curaçao Bonaire

ST VINCENT AND THE GRENADINES 1979

GRENADA 1974 Invaded 1983

VENEZUELA 1830

ST. LUCIA 1979

BARBADOS 1966

TRINIDAD AND TOBAGO 1962

GUYANA 1966

1920

COPYRIGHT PHILIP'S

DENSITY of POPULATION

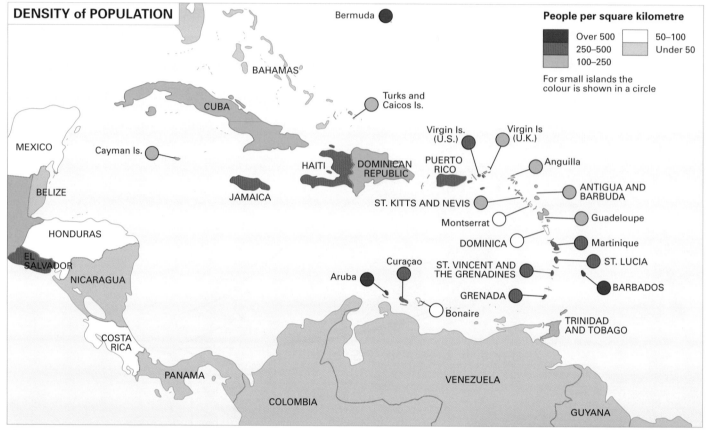

People per square kilometre

■	Over 500	□	50–100
	250–500		Under 50
	100–250		

For small islands the colour is shown in a circle

Bermuda

BAHAMAS
CUBA
Turks and Caicos Is.
MEXICO
Cayman Is.
HAITI DOMINICAN REPUBLIC
PUERTO RICO
Virgin Is. (U.S.) Virgin Is (U.K.)
Anguilla
BELIZE
JAMAICA
ST. KITTS AND NEVIS
ANTIGUA AND BARBUDA
Montserrat
Guadeloupe
HONDURAS
DOMINICA
Martinique
EL SALVADOR
ST. LUCIA
NICARAGUA
Curaçao ST. VINCENT AND THE GRENADINES
Aruba
BARBADOS
GRENADA
Bonaire
TRINIDAD AND TOBAGO
COSTA RICA
PANAMA
VENEZUELA
COLOMBIA
GUYANA

The Caribbean has both areas of high and low population density. Port of Spain, Trinidad (left) is high density, yet to its south you can see unoccupied swamplands. Guyana is a large country with a small population. The Shulinab Amerindian Community in Rupununi, Guyana (right) is typical of an area of low population density.

LIFE EXPECTANCY

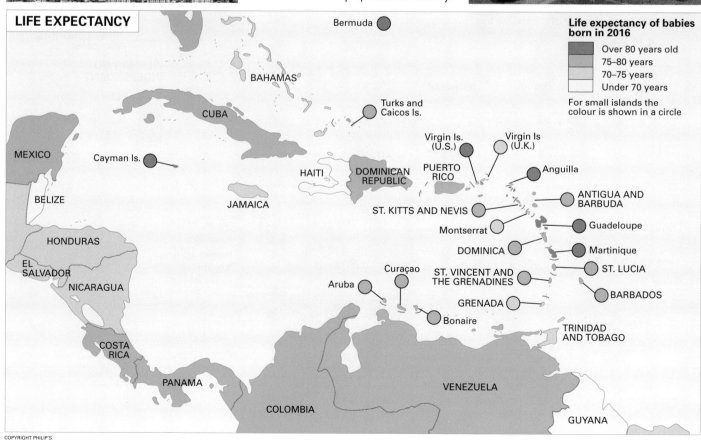

Life expectancy of babies born in 2016

■	Over 80 years old
	75–80 years
	70–75 years
□	Under 70 years

For small islands the colour is shown in a circle

Bermuda

BAHAMAS
CUBA
Turks and Caicos Is.
MEXICO
Cayman Is.
HAITI DOMINICAN REPUBLIC
PUERTO RICO
Virgin Is. (U.S.) Virgin Is (U.K.)
Anguilla
BELIZE
JAMAICA
ST. KITTS AND NEVIS
ANTIGUA AND BARBUDA
Montserrat
Guadeloupe
HONDURAS
DOMINICA
Martinique
EL SALVADOR
ST. LUCIA
NICARAGUA
Curaçao ST. VINCENT AND THE GRENADINES
Aruba
GRENADA
Bonaire
BARBADOS
TRINIDAD AND TOBAGO
COSTA RICA
PANAMA
VENEZUELA
COLOMBIA
GUYANA

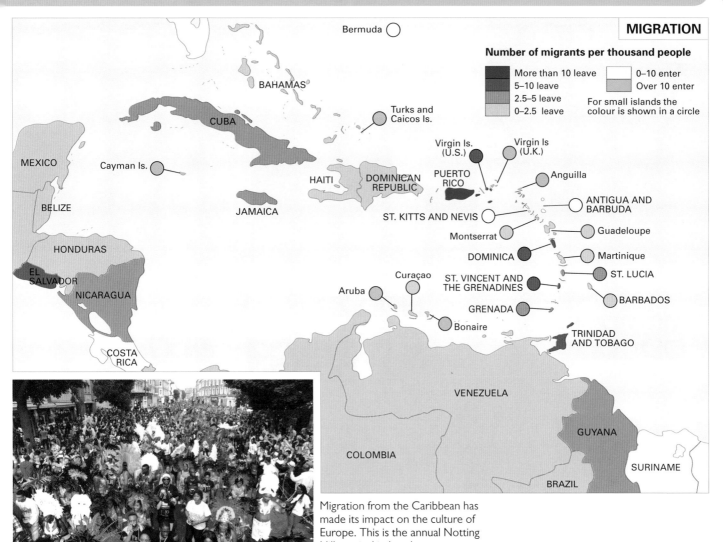

MIGRATION

Bermuda

Number of migrants per thousand people

More than 10 leave	0–10 enter
5–10 leave	Over 10 enter
2.5–5 leave	For small islands the
0–2.5 leave	colour is shown in a circle

BAHAMAS

Turks and Caicos Is.

CUBA

MEXICO

Cayman Is.

Virgin Is. (U.S.)

Virgin Is (U.K.)

HAITI

DOMINICAN REPUBLIC

PUERTO RICO

Anguilla

ANTIGUA AND BARBUDA

BELIZE

JAMAICA

ST. KITTS AND NEVIS

Montserrat

Guadeloupe

HONDURAS

DOMINICA

Martinique

EL SALVADOR

NICARAGUA

Curaçao

Aruba

ST. VINCENT AND THE GRENADINES

ST. LUCIA

BARBADOS

GRENADA

Bonaire

COSTA RICA

TRINIDAD AND TOBAGO

VENEZUELA

GUYANA

COLOMBIA

SURINAME

BRAZIL

Migration from the Caribbean has made its impact on the culture of Europe. This is the annual Notting Hill carnival in London.

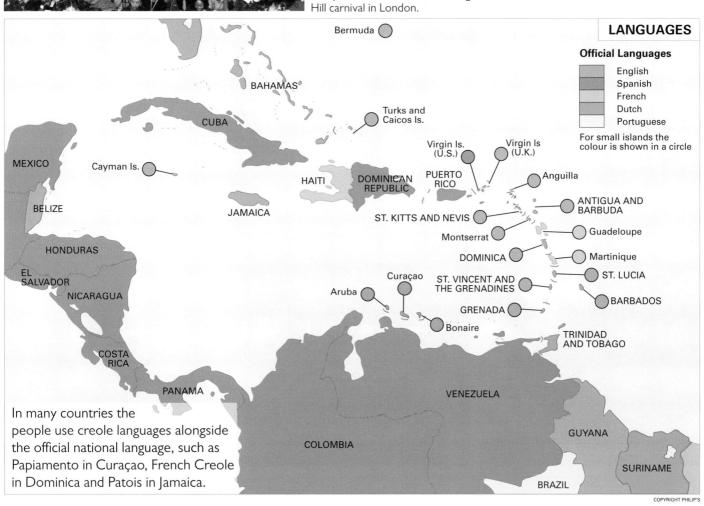

LANGUAGES

Bermuda

Official Languages

	English
	Spanish
	French
	Dutch
	Portuguese

For small islands the colour is shown in a circle

BAHAMAS

Turks and Caicos Is.

CUBA

MEXICO

Cayman Is.

Virgin Is. (U.S.)

Virgin Is (U.K.)

HAITI

DOMINICAN REPUBLIC

PUERTO RICO

Anguilla

ANTIGUA AND BARBUDA

BELIZE

JAMAICA

ST. KITTS AND NEVIS

Montserrat

Guadeloupe

HONDURAS

DOMINICA

Martinique

EL SALVADOR

NICARAGUA

Curaçao

Aruba

ST. VINCENT AND THE GRENADINES

ST. LUCIA

BARBADOS

GRENADA

Bonaire

COSTA RICA

TRINIDAD AND TOBAGO

PANAMA

VENEZUELA

In many countries the people use creole languages alongside the official national language, such as Papiamento in Curaçao, French Creole in Dominica and Patois in Jamaica.

GUYANA

COLOMBIA

SURINAME

BRAZIL

Citrus farms in Belize produce oranges on a large scale for processing to juice.

Jamaica's Blue Mountain coffee beans are handpicked to preserve quality.

Jamaican bananas for the export market have to be carefully prepared before shipping.

AGRICULTURAL PRODUCTION

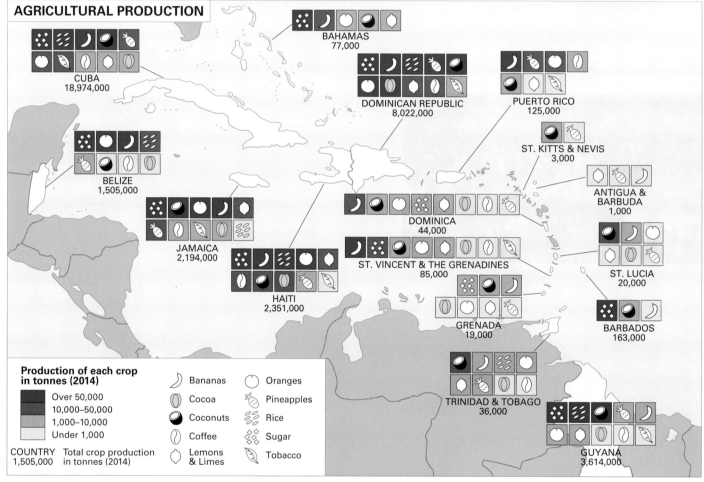

BAHAMAS
77,000

CUBA
18,974,000

DOMINICAN REPUBLIC
8,022,000

PUERTO RICO
125,000

ST. KITTS & NEVIS
3,000

BELIZE
1,505,000

ANTIGUA & BARBUDA
1,000

JAMAICA
2,194,000

DOMINICA
44,000

ST. LUCIA
20,000

ST. VINCENT & THE GRENADINES
85,000

HAITI
2,351,000

GRENADA
19,000

BARBADOS
163,000

TRINIDAD & TOBAGO
36,000

GUYANA
3,614,000

Production of each crop in tonnes (2014)

Over 50,000
10,000–50,000
1,000–10,000
Under 1,000

COUNTRY   Total crop production
1,505,000   in tonnes (2014)

Bananas
Cocoa
Coconuts
Coffee
Lemons & Limes

Oranges
Pineapples
Rice
Sugar
Tobacco

Sugar cane is increasingly harvested by machine, as it is in St. Kitts.
COPYRIGHT PHILIP'S

Grenada is the world's second largest producer of nutmeg. Nutmeg products are exported.

Machines are now used to harvest the rice crop along Guyana's flat coastal lands.

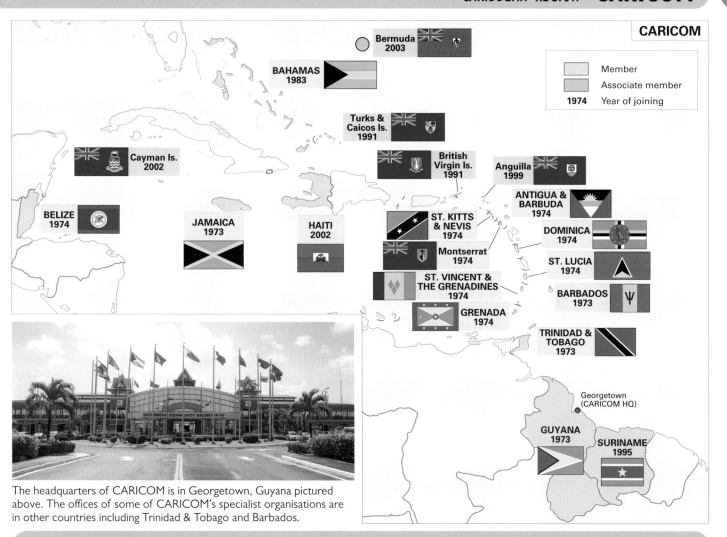

CARICOM

Bermuda 2003

BAHAMAS 1983

Turks & Caicos Is. 1991

Cayman Is. 2002

British Virgin Is. 1991

Anguilla 1999

ANTIGUA & BARBUDA 1974

BELIZE 1974

JAMAICA 1973

HAITI 2002

ST. KITTS & NEVIS 1974

DOMINICA 1974

Montserrat 1974

ST. LUCIA 1974

ST. VINCENT & THE GRENADINES 1974

BARBADOS 1973

GRENADA 1974

TRINIDAD & TOBAGO 1973

Georgetown (CARICOM HQ)

GUYANA 1973

SURINAME 1995

Member
Associate member
1974 Year of joining

The headquarters of CARICOM is in Georgetown, Guyana pictured above. The offices of some of CARICOM's specialist organisations are in other countries including Trinidad & Tobago and Barbados.

ORGANISATIONS OF THE CARIBBEAN COMMUNITY

Caribbean Agricultural Development Institute - CARDI
Caribbean Agricultural Health and Safety Agency - CAHFSA
Caribbean Aviation Safety and Security Oversight System - CASSOS
Caribbean Centre for Renewable Energy and Energy Efficiency - CCREEE
Caribbean Centre for Development Administration - CARICAD
Caribbean Community Climate Change Centre - CCCCC
Caribbean Court of Justice - CJC
Caribbean Examinations Council - CXC
Caribbean Institute for Meteorology and Hydrology - CIMHH
Caribbean Meteorological Organisation - CMO

Caribbean Regional Fisheries Mechanism - CRFM
Caribbean Telecommunications Union - CTU
Caribbean Competition Commission - CCC
Caribbean Development Fund - CDF
Caribbean Implementing Agency for Crime and Security - IMPACS

Examples of associate institutions
Caribbean Development Bank - CDB
Caribbean Disaster Emergency Management Agency - CDEMA
Caribbean Export Development Agency - Carib-Export
Caribbean Law Institute - CLI
Caribbean Tourism Organisation - CTO

The Caribbean Community has established many organisations to improve the region's economy, safety and justice. These are listed above. Other regional organisations are associated with CARICOM. Some examples of these are listed.

- Using the name for each organisation, can you work out how it will help the people of the Caribbean?
- Most organisations have a logo to give it identity. Can you match the logos below with one of the organisations listed?

CARICOM

The Caribbean Community (CARICOM) began in 1973 when Barbados, Guyana, Jamaica and Trinidad and Tobago signed the Treaty of Chaguaramas (in Trinidad). The other members joined in 1974 and at later dates with Suriname the last to join in 1995. The aims are: economic cooperation within the Caribbean Common Market, coordination of foreign policy and cooperation in the areas of health, education, technology, transport, culture and sport. The headquarters is in Guyana.

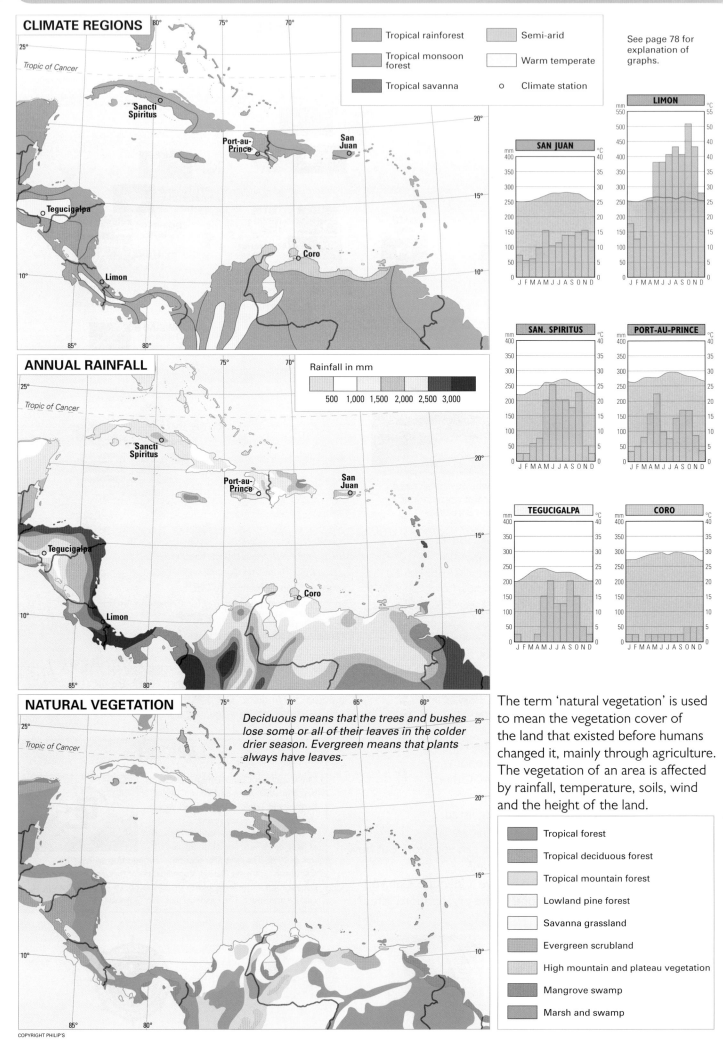

CLIMATE REGIONS

Tropical rainforest
Tropical monsoon forest
Tropical savanna
Semi-arid
Warm temperate
○ Climate station

See page 78 for explanation of graphs.

Tropic of Cancer

Sancti Spiritus
Port-au-Prince
San Juan
Tegucigalpa
Coro
Limon

ANNUAL RAINFALL

Rainfall in mm

500 1,000 1,500 2,000 2,500 3,000

Tropic of Cancer

Sancti Spiritus
Port-au-Prince
San Juan
Tegucigalpa
Coro
Limon

LIMON
SAN JUAN
SAN. SPIRITUS
PORT-AU-PRINCE
TEGUCIGALPA
CORO

NATURAL VEGETATION

Tropic of Cancer

Deciduous means that the trees and bushes lose some or all of their leaves in the colder drier season. Evergreen means that plants always have leaves.

The term 'natural vegetation' is used to mean the vegetation cover of the land that existed before humans changed it, mainly through agriculture. The vegetation of an area is affected by rainfall, temperature, soils, wind and the height of the land.

Tropical forest
Tropical deciduous forest
Tropical mountain forest
Lowland pine forest
Savanna grassland
Evergreen scrubland
High mountain and plateau vegetation
Mangrove swamp
Marsh and swamp

TOURIST ARRIVALS

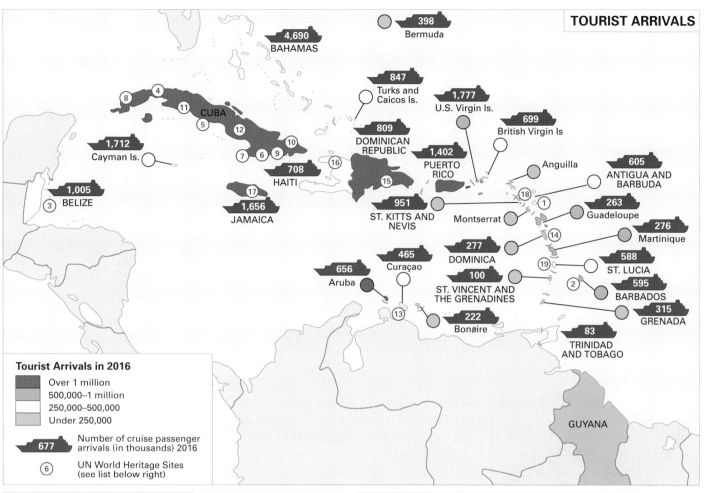

- 398 Bermuda
- 4,690 BAHAMAS
- 847 Turks and Caicos Is.
- 1,777 U.S. Virgin Is.
- 699 British Virgin Is
- 809 DOMINICAN REPUBLIC
- 1,402 PUERTO RICO
- 605 ANTIGUA AND BARBUDA
- Anguilla
- 1,712 Cayman Is.
- 708 HAITI
- 951 ST. KITTS AND NEVIS
- Montserrat
- 263 Guadeloupe
- 276 Martinique
- 1,005 BELIZE
- 1,656 JAMAICA
- 465 Curaçao
- 277 DOMINICA
- 588 ST. LUCIA
- 595 BARBADOS
- 656 Aruba
- 100 ST. VINCENT AND THE GRENADINES
- 222 Bonaire
- 315 GRENADA
- 83 TRINIDAD AND TOBAGO
- GUYANA

Tourist Arrivals in 2016

- Over 1 million
- 500,000–1 million
- 250,000–500,000
- Under 250,000

677 Number of cruise passenger arrivals (in thousands) 2016

6 UN World Heritage Sites (see list below right)

EMPLOYMENT IN TOURISM

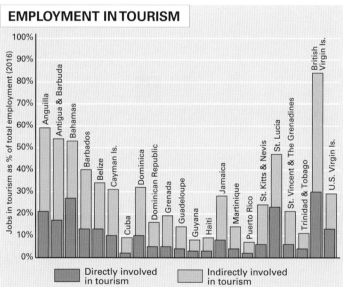

Jobs in tourism as % of total employment (2016)

Countries listed: Anguilla, Antigua & Barbuda, Bahamas, Barbados, Belize, Cayman Is., Cuba, Dominica, Dominican Republic, Grenada, Guadeloupe, Guyana, Haiti, Jamaica, Martinique, Puerto Rico, St. Kitts & Nevis, St. Lucia, St. Vincent & The Grenadines, Trinidad & Tobago, British Virgin Is., U.S. Virgin Is.

- Directly involved in tourism
- Indirectly involved in tourism

UN WORLD HERITAGE SITES IN THE CARIBBEAN

1. Antigua Naval Dockyard, Antigua & Barbuda
2. Historic Bridgetown and its Garrison, Barbados
3. Belize Barrier Reef Reserve System, Belize
4. Old Havana and its Fortifications, Cuba
5. Trinidad and the Valley de los Ingenios, Cuba
6. San Pedro de la Roca Castle, Santiago de Cuba, Cuba
7. Desembarco del Granma National Park, Cuba
8. Viñales Valley, Cuba
9. Landscapes of the First Coffee Plantations, Cuba
10. Alejandro de Humboldt National Park, Cuba
11. Historic Centre of Cienfuegos, Cuba
12. Historic Centre of Camagüey, Cuba
13. Historic Area of Willemstad, Inner City and Harbour, Curaçao
14. Morne Trois Pitons National Park, Dominica
15. Colonial City of Santo Domingo, Dominican Republic
16. National History Park, Sans Souci Citadel, Haiti
17. Blue and John Crow Mountains, Jamaica
18. Brimstone Hill Fortress National Park, St. Kitts & Nevis
19. Pitons Management Area, St. Lucia

IMPORTANCE OF TOURISM

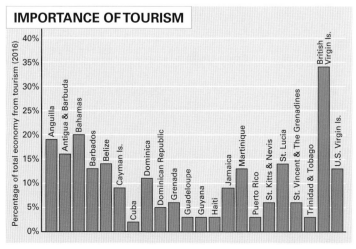

Percentage of total economy from tourism (2016)

Countries listed: Anguilla, Antigua & Barbuda, Bahamas, Barbados, Belize, Cayman Is., Cuba, Dominica, Dominican Republic, Grenada, Guadeloupe, Guyana, Haiti, Jamaica, Martinique, Puerto Rico, St. Kitts & Nevis, St. Lucia, St. Vincent & The Grenadines, Trinidad & Tobago, British Virgin Is., U.S. Virgin Is.

Selected by UNESCO, World Heritage sites are places of great international importance. The Pitons in St. Lucia (two volcanic mountains) are a good example of such a site.

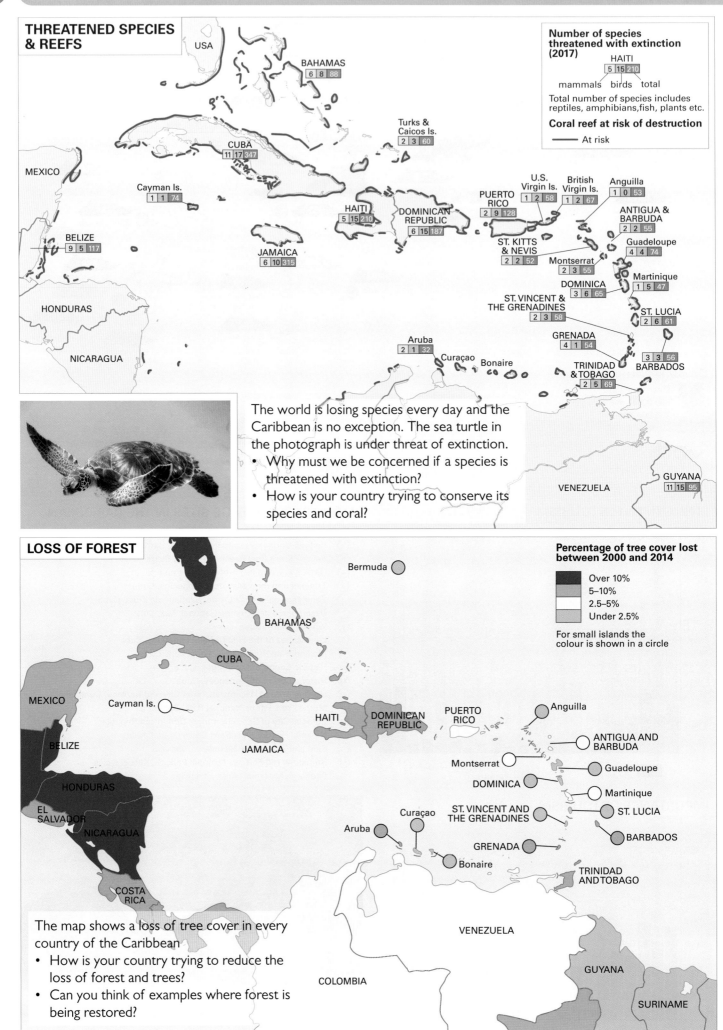

THREATENED SPECIES & REEFS

Number of species threatened with extinction (2017)

HAITI
| 5 | 15 | 210 |

mammals birds total

Total number of species includes reptiles, amphibians, fish, plants etc.

Coral reef at risk of destruction

—— At risk

USA

BAHAMAS
| 6 | 8 | 88 |

Turks & Caicos Is.
| 2 | 3 | 60 |

CUBA
| 11 | 17 | 347 |

MEXICO

Cayman Is.
| 1 | 1 | 74 |

HAITI
| 5 | 15 | 210 |

DOMINICAN REPUBLIC
| 6 | 15 | 187 |

PUERTO RICO
| 2 | 9 | 128 |

U.S. Virgin Is.
| 1 | 2 | 58 |

British Virgin Is.
| 1 | 2 | 67 |

Anguilla
| 1 | 0 | 53 |

ANTIGUA & BARBUDA
| 2 | 2 | 55 |

Guadeloupe
| 4 | 4 | 74 |

ST. KITTS & NEVIS
| 2 | 2 | 52 |

Montserrat
| 2 | 3 | 55 |

Martinique
| 1 | 5 | 47 |

BELIZE
| 9 | 5 | 117 |

JAMAICA
| 6 | 10 | 315 |

DOMINICA
| 3 | 6 | 65 |

ST. VINCENT & THE GRENADINES
| 2 | 3 | 58 |

ST. LUCIA
| 2 | 6 | 61 |

HONDURAS

NICARAGUA

Aruba
| 2 | 1 | 32 |

Curaçao Bonaire

GRENADA
| 4 | 1 | 54 |

TRINIDAD & TOBAGO
| 2 | 5 | 69 |

BARBADOS
| 3 | 3 | 56 |

VENEZUELA

GUYANA
| 11 | 15 | 95 |

The world is losing species every day and the Caribbean is no exception. The sea turtle in the photograph is under threat of extinction.
- Why must we be concerned if a species is threatened with extinction?
- How is your country trying to conserve its species and coral?

LOSS OF FOREST

Percentage of tree cover lost between 2000 and 2014

- Over 10%
- 5–10%
- 2.5–5%
- Under 2.5%

For small islands the colour is shown in a circle

Bermuda

BAHAMAS

CUBA

MEXICO

Cayman Is.

HAITI DOMINICAN REPUBLIC PUERTO RICO

Anguilla

BELIZE

JAMAICA

ANTIGUA AND BARBUDA

Montserrat

Guadeloupe

HONDURAS

DOMINICA

Martinique

EL SALVADOR

NICARAGUA

Aruba Curaçao Bonaire

ST. VINCENT AND THE GRENADINES

GRENADA

ST. LUCIA

BARBADOS

COSTA RICA

TRINIDAD AND TOBAGO

The map shows a loss of tree cover in every country of the Caribbean
- How is your country trying to reduce the loss of forest and trees?
- Can you think of examples where forest is being restored?

VENEZUELA

COLOMBIA

GUYANA

SURINAME

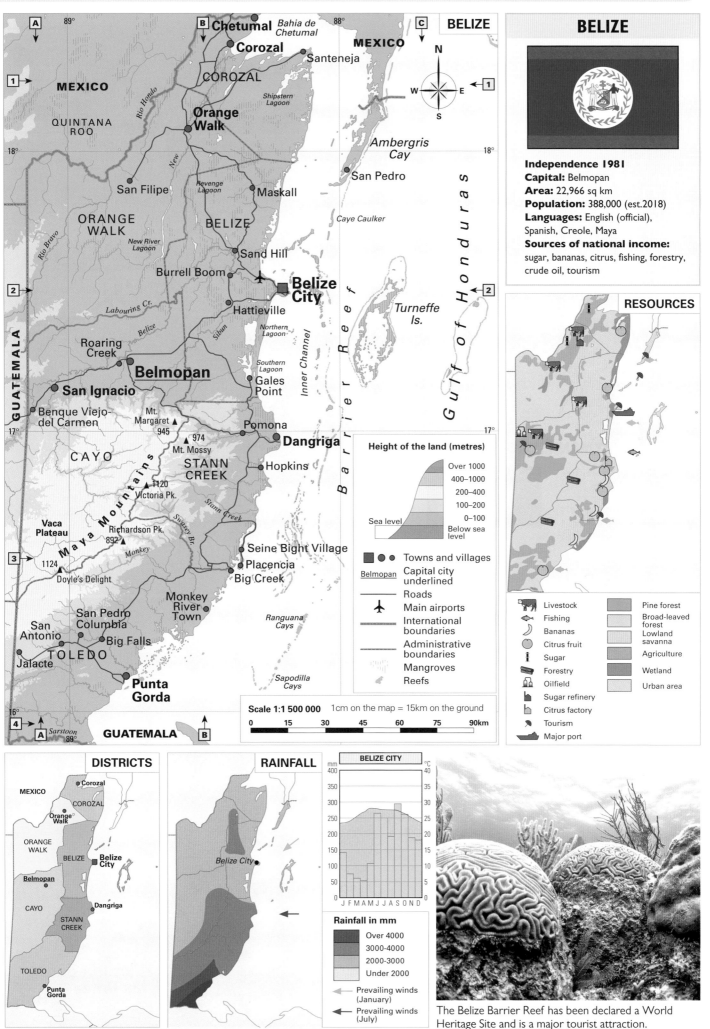

BELIZE

BELIZE

Independence 1981
Capital: Belmopan
Area: 22,966 sq km
Population: 388,000 (est.2018)
Languages: English (official),
Spanish, Creole, Maya
Sources of national income:
sugar, bananas, citrus, fishing, forestry,
crude oil, tourism

RESOURCES

Livestock
Fishing
Bananas
Citrus fruit
Sugar
Forestry
Oilfield
Sugar refinery
Citrus factory
Tourism
Major port

Pine forest
Broad-leaved
forest
Lowland
savanna
Agriculture
Wetland
Urban area

Height of the land (metres)

Over 1000
400–1000
200–400
100–200
0–100
Sea level
Below sea
level

■ ● ● Towns and villages
Belmopan Capital city
underlined
Roads
Main airports
International
boundaries
Administrative
boundaries
Mangroves
Reefs

Scale 1:1 500 000 1cm on the map = 15km on the ground
0 15 30 45 60 75 90km

DISTRICTS

MEXICO
Corozal
COROZAL
Orange
Walk
ORANGE
WALK
BELIZE
Belize
City
Belmopan
CAYO
Dangriga
STANN
CREEK
TOLEDO
Punta
Gorda

RAINFALL

Belize City

Rainfall in mm
Over 4000
3000-4000
2000-3000
Under 2000
← Prevailing winds
(January)
← Prevailing winds
(July)

BELIZE CITY

The Belize Barrier Reef has been declared a World
Heritage Site and is a major tourist attraction.

COPYRIGHT PHILIP'S

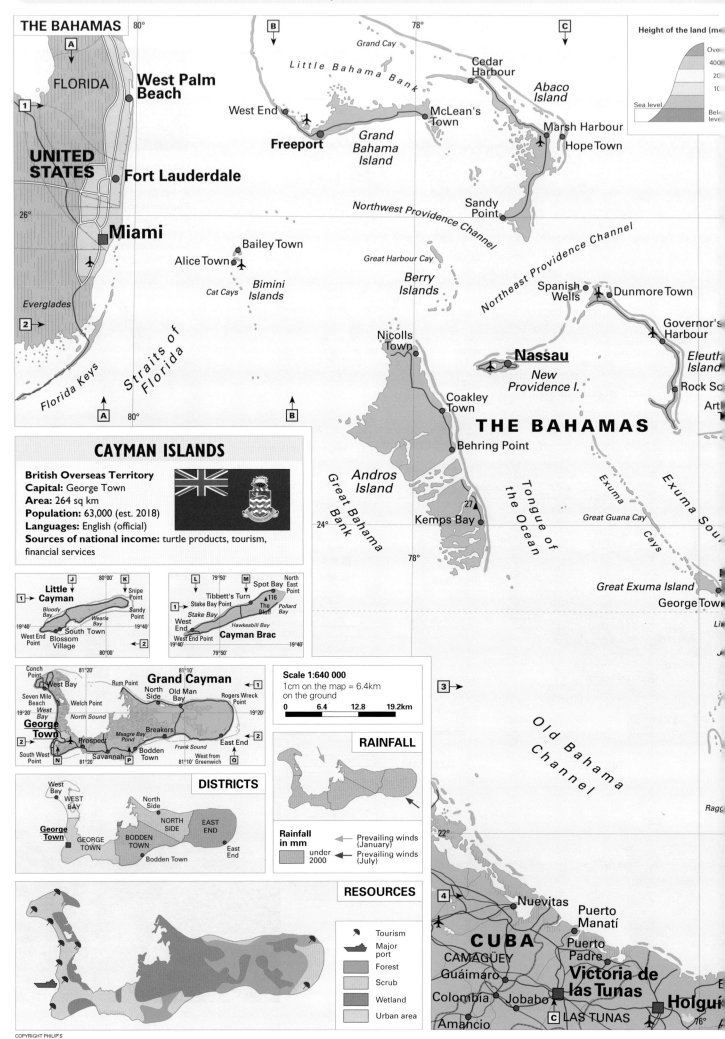

THE BAHAMAS

80°   78°

FLORIDA

West Palm Beach

West End

Freeport

Grand Bahama Island

Little Bahama Bank

Grand Cay

Cedar Harbour

Abaco Island

McLean's Town

Marsh Harbour

Hope Town

UNITED STATES

Fort Lauderdale

26°

Miami

Bailey Town

Alice Town

Bimini Islands

Cat Cays

Everglades

Straits of Florida

Florida Keys

Sandy Point

Northwest Providence Channel

Great Harbour Cay

Berry Islands

Northeast Providence Channel

Spanish Wells

Dunmore Town

Governor's Harbour

Nicolls Town

Nassau

New Providence I.

Eleuth Island

Rock Sc

Art

Coakley Town

THE BAHAMAS

Behring Point

Andros Island

Great Bahama Bank

Kemps Bay

24°

27

Tongue of the Ocean

Exuma Cays

Exuma Sou

Great Guana Cay

78°

Great Exuma Island

George Tow

Li

J

Height of the land (me
Ove
400
20
10
Sea level
Belo
leve

CAYMAN ISLANDS

British Overseas Territory
Capital: George Town
Area: 264 sq km
Population: 63,000 (est. 2018)
Languages: English (official)
Sources of national income: turtle products, tourism, financial services

Little Cayman

80°00'
Snipe Point
Bloody Bay
Wearis Bay
Sandy Point
West End Point
South Town
Blossom Village
19°40'

Cayman Brac

79°50'
Tibbett's Turn
Spot Bay
North East Point
Stake Bay Point
Stake Bay
▲116
The Bluff
Pollard Bay
West End
Hawkesbill Bay
West End Point
Cayman Brac
19°40'

Grand Cayman

Conch Point
81°20'
West Bay
Rum Point
81°10'
Seven Mile Beach
North Side
Old Man Bay
Rogers Wreck Point
19°20'
Welch Point
West Bay
North Sound
George Town
Breakers
Meagre Bay Pond
East End
Prospect
Savannah
Bodden Town
Frank Sound
West from Greenwich
South West Point
81°20'
81°10'
19°20'

Scale 1:640 000
1cm on the map = 6.4km on the ground
0   6.4   12.8   19.2km

DISTRICTS

West Bay
WEST BAY
North Side
NORTH SIDE
EAST END
George Town
GEORGE TOWN
BODDEN TOWN
East End
Bodden Town

RAINFALL

Rainfall in mm
under 2000
Prevailing winds (January)
Prevailing winds (July)

RESOURCES

Tourism
Major port
Forest
Scrub
Wetland
Urban area

Old Bahama Channel

22°

Ragg

Nuevitas

Puerto Manatí

CUBA

CAMAGÜEY

Puerto Padre

Guáimaro

Victoria de las Tunas

Colombia

Jobabo

Holguí

Amancio

LAS TUNAS

76°

Legend:
Urban areas
Towns and villages
Capital city underlined
Highways
Roads
Main airports
International boundaries
Mangroves   Reefs

TURKS & CAICOS ISLANDS

22°
G
North Caicos
Caicos Passage
Bottle Creek
Providenciales I.
Conch Bar
Bambarra
Blue Hills
West Caicos I.
Middle Caicos
C a i c o s
I s l a n d s
East Caicos
35
26°
South Caicos
Cockburn Harbour
Caicos Bank
1
Ambergris Cays
Seal Cays
G
West 72° from Greenwich
H

N
E
S

H

TURKS & CAICOS ISLANDS

British Overseas Territory
Capital: Cockburn Town
Area: 430 sq km
Population: 40,000 (est.2018)
Languages: English (official)
Sources of national income: fishing, tourism, financial services

ATLANTIC OCEAN

Cockburn Town
Grand Turk I.
Turks Island Passage
Salt Cay
Turks Islands
1

Scale 1:1 240 000
1cm on the map = 12.4km on the ground
0   12.4   24.8   37.2   49.6km

74°
E
72°
F

THE COMMONWEALTH OF THE BAHAMAS

Independence 1973
Capital: Nassau
Area: 13,878 sq km
Population: 351,000 (2010 census)
Languages: English (official), Creole
Sources of national income:
fishing, tourism, international banking

2

This cruise ship is leaving Nassau and heading to its next stop in the Dominican Republic. More than half of The Bahamas' five million tourists each year are cruise passengers. Tourism is key to the country's economy.

24°

nd
Bight
San Salvador I.
Cockburn Town
Rum Cay
Long Island
Tropic of Cancer
Deadman's Cay
rence Town
Crooked I.
Samana Cay
Colonel Hill
Plana Cays
Albert Town
Mayaguana Passage
Mayaguana I.
Abraham's Bay
Acklins I.

3

ATLANTIC OCEAN

Caicos Passage
See map above

North Caicos
Middle Caicos
East Caicos
Providenciales
Caicos Islands
Turks & Caicos Is.
(U.K.)
Cockburn Town
Turks Islands

22°

Little Inagua Island

Scale 1:2 700 000
m on the map = 27km on the ground
27   54   81   108   135km

Matthew Town
Lake Rose
Great Inagua Island

D
74°
E
72°
F

4

COPYRIGHT PHILIP'S

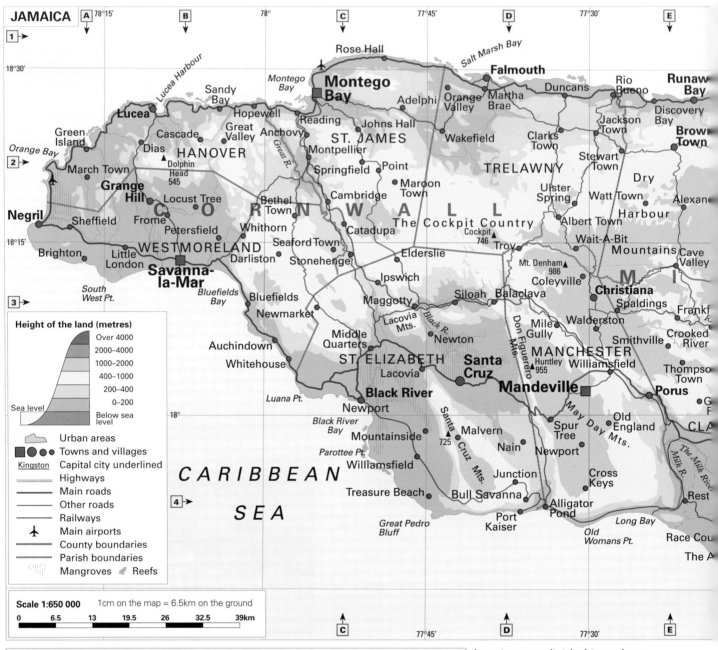

JAMAICA

78°15' · 78° · 77°45' · 77°30'

18°30'
18°15'
18°

Rose Hall · Salt Marsh Bay · Falmouth · Rio Bueno · Runaw Bay
Montego Bay · Montego Bay · Adelphi · Orange Valley · Martha Brae · Duncans · Jackson Town · Discovery Bay · Brown Town
Lucea Harbour · Sandy Bay · Hopewell · Reading · Johns Hall · Wakefield · Clarks Town · Stewart Town · Dry
Lucea · Great Valley · Anchovy · ST. JAMES · Ulster Spring · Watt Town · Alexan
Green Island · Cascade · Montpellier · TRELAWNY · Albert Town · Harbour
Dias · HANOVER · Springfield · Point · Maroon Town · The Cockpit Country · Wait-A-Bit
Dolphin Head 545 · Cambridge · Cockpit 746 · Mountains · Cave Valley
March Town · Grange Hill · Locust Tree · Bethel Town · Catadupa · Troy
Negril · Sheffield · Frome · Whithorn · Seaford Town · Elderslie · Mt. Denham 986 · Coleyville · Christiana · Spaldings · Frankf
Petersfield · WESTMORELAND · Stonehenge · Ipswich · Siloah · Balaclava · Walderston · Crooked River
Brighton · Little London · Darliston · Maggotty · Mile Gully · Smithville
Savanna-la-Mar · Bluefields Bay · Bluefields · Lacovia Mts. · Newton · MANCHESTER · Williamsfield · Thompso Town
South West Pt. · Newmarket · Middle Quarters · ST. ELIZABETH · Santa Cruz · Huntley 955 · Porus · G
Auchindown · Whitehouse · Lacovia · Mandeville · Old England · CLA
Luana Pt. · Black River · Newport · Spur Tree · Newport
Black River Bay · Mountainside · Malvern · May Day Mts.
Parottee Pt. · Williamsfield · Nain · Cross Keys · The Milk Rive
Treasure Beach · Junction · Rest
Great Pedro Bluff · Bull Savanna · Alligator Pond · Long Bay · Race Cou
Port Kaiser · Old Womans Pt. · The A

CARIBBEAN SEA

Height of the land (metres)
Over 4000
2000–4000
1000–2000
400–1000
200–400
0–200
Sea level
Below sea level

Urban areas
Towns and villages
Kingston Capital city underlined
Highways
Main roads
Other roads
Railways
Main airports
County boundaries
Parish boundaries
Mangroves · Reefs

Scale 1:650 000 · 1cm on the map = 6.5km on the ground
0 · 6.5 · 13 · 19.5 · 26 · 32.5 · 39km

PARISHES

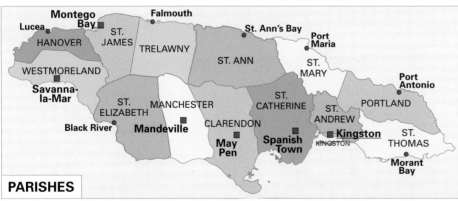

Montego Bay · Falmouth · St. Ann's Bay · Port Maria
Lucea · ST. JAMES · TRELAWNY · ST. ANN · ST. MARY
HANOVER · Port Antonio
WESTMORELAND · ST. CATHERINE · PORTLAND
Savanna-la-Mar · ST. ELIZABETH · MANCHESTER · ST. ANDREW
Black River · Mandeville · CLARENDON · Kingston · ST. THOMAS
May Pen · Spanish Town · KINGSTON · Morant Bay

Jamaica was divided into three counties, Cornwall, Middlesex and Surrey, in 1758 by the British governor. The counties now have little administrative function.

Kingston became the capital of Jamaica in 1872. Before this, the capital was Spanish Town. Jamaica is divided into 14 parishes, although Kingston and St. Andrew parishes are jointly administered.

COUNTIES

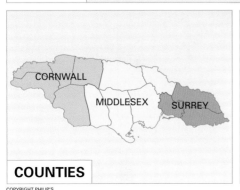

CORNWALL · MIDDLESEX · SURREY

NATIONAL SYMBOLS & COAT OF ARMS

Ackee, national fruit

Blue Mahoe, national tree

The Doctor Bird, national bird

Lignum Vitae, national flower

Coat of Arms

COPYRIGHT PHILIP'S

F ↓ 77° G ↓

N
W · E
S

← 1

18°30'

← 2

JAMAICA

Independence 1962
Capital: Kingston
Area: 10, 990 sq km
Population: 2,698,000 (2011 census)
Languages: English, Jamaican Creole
Sources of national income:
sugar, chemicals, coffee, tourism,
alumina, bauxite, clothing

Renewal of downtown Kingston looking
north from the Harbour

76°30'

H ↓

Drax Hall Pt.
St. Ann's Bay
mboo
Epworth   **Ocho Rios**   Galina Pt.
emont   Oracabessa
Albion 841 Golden Grove   Gayle   **Port Maria**
ST. ANN   Moneague   **Highgate**   **ST. MARY**   Don Christophers Pt.
Bensonton   Annotto Bay
                 Richmond   **Annotto Bay**   Palmetto Pt.
D **L** **E** **S** **E** **X**   Buff Bay   Buff Bay   18°15'
ead   Troja   Savanna Pt.
Ewarton   Riversdale   Castleton   Hope Bay   **St. Margaret's Bay**
Linstead   Glengoffe   **Port Antonio**   North East Pt.   76°15'
Crofts Hill Lluidas Vale **Bog Walk** Sligoville Section   Fellowship   Boston Bay   ← 3
apelton   **ST.**   Cooper's Hill   Stony   Catherines Pk. **PORTLAND**   Long Bay
Rock River Point Hill   779▲   Hill 1541▲ **Blue Mountains**   Millbank
CATHERINE Guanaboa Caymanas Constant   2256▲   Macca
            Vale   Spring **ST. ANDREW** Blue Mountain Pk.   Sucker   Happy
Spanish Town   **Half Way** **S** Mavis 1336 Grove   18°
ON   **Tree**   Bank Cedar Valley **ST. THOMAS**
May Pen **Portmore** **KINGSTON**   Ramble   Holland
            Braeton   Kingston Harbour Trinity Ville Bath Bay
Freetown   Harbour Bull   Seaforth   Golden
Port   **Old Harbour** Port View Bay Easington **Port** Grove
Esquivel   **Old Harbour** Royal The   **Morant** Morant Morant Pt.
Hayes   **Bay**   Palisadoes   **Bay**   ← 4
alt River   Cabarita Pt.   Cow Bay Pt. **Yallahs** White
Lionel   Rocky Pt.   Yallahs Pt.   Horses
Town   Portland Bight

C A R I B B E A N

17°45'

S E A   ← 5

Portland Pt. F   77°   G   76°45'   H   76°30'   J   76°15'

Urban areas

Parks/recreation
areas/woodland

Towns and villages

Highways

Main roads

Other roads

Railways

KINGSTON & PORTMORE

NEW HAVEN   18°   MOLYNES GARDENS   CAMPERDOWN   GRANTS PEN   HOPE PASTURES 18°

Six Miles   BARBICAN

Fresh River   Sandy Gully   MOLYNES SQUARE   **LIGUANEA PLAIN**   Liguanea   MONA HEIGHTS   **Papine**

OLYMPIC GARDENS   **Half Way Tree**   University of the West Indies

FOUR MILE   COCKBURN GARDENS   **New Kingston**   BEVERLY HILLS   Mona Reservoir

GREGORY PARK   RICHMOND PARK   KENCOT   INDEPENDENCE PARK   AUGUST TOWN

Caymanas Park Race Course   **Three Mile**   Tinson Pen Airstrip   WHITFIELD TOWN   **Cross Roads**   EDEN GARDENS   Hope River

WATERFORD   NEWPORT WEST   MOUNTAIN VIEW GARDENS

Rio Cobre   Hunts Bay   GREENWICH TOWN   TRENCH TOWN   NATIONAL HEROES PARK   ROLLINGTON TOWN   NORMAN GARDENS   Long Mountain

INDEPENDENCE CITY   CUMBERLAND   Saint William Grant Park   BOURNEMOUTH GARDENS

PORTMORE   WESTMEADE   Dawkins Lagoon   FORT AUGUSTA   NEWPORT EAST

PORTMORE PINES   BRIDGEPORT   **KINGSTON**   **Harbour View**

DAYTONA   BRAETON   PORT HENDERSON   Kingston Harbour   Harbour Head

GREATER PORTMORE   Norman Manley Highway

REFUGE CAY

Port Royal
PORT ROYAL POINT   _The Palisadoes_   NORMAN MANLEY INTERNATIONAL AIRPORT

76°45'

Scale 1:100 000
1cm on the map = 1km on the ground
0 1 2 3 4km

COPYRIGHT PHILIP'S

PARISHES

HANOVER

Capital: Lucea
Area: 451 sq km
Population: 70,000 (2011 census)
Main towns: Green Island, Hopewell, Sandy Bay
Farming: yam, sugar cane, cattle

Profile: Hanover, in the north-west of the island, is Jamaica's second smallest parish. Lucea (once known as St. Lucia) grew around a safe harbour and was once a thriving port. At the entrance to Lucea Harbour stands Fort Charlotte, constructed in 1761. The economic focus of coastal areas and river valleys is agriculture. Tourism is also important, especially around Hopewell.

ST. JAMES

Capital: Montego Bay
Area: 595 sq km
Population: 185,000 (2011 census)
Main towns: Anchovy, Cambridge, Montpellier
Farming: sugar cane, bananas, cattle, forestry

Profile: Montego Bay is the largest city in the west of Jamaica. The banana trade brought prosperity to Montego Bay in the 19th century. Tourism grew and is now the main employer. Sangster International Airport and Montego Bay's cruise ship piers are the busiest in Jamaica. Manufacturing is also important, especially clothing, data entry and food processing.

TRELAWNY

Capital: Falmouth
Area: 874 sq km
Population: 76,000 (2011 census)
Main towns: Martha Brae, Rio Bueno, Stewart Town, Wait-a-Bit, Duncans
Farming: bananas, sugar cane, yam, coffee, fruit, vegetables

Profile: The north of the parish is flat with the wide Queen of Spain valley inland. Trelawny has more sugar estates than any other parish and sugar and rum are still the most important products. Fishing is important along the north coast. Much of the southern half of the parish is the uninhabited, but wildlife rich, Cockpit Country.

WESTMORELAND

Capital: Savanna-la-Mar
Area: 785 sq km
Population: 145,000 (2011 census)
Main towns: Bethel Town, Bluefields, Grange Hill, Negril, Seaford Town
Farming: sugar cane, bananas, dairy cattle

Profile: With fertile plains and many rivers, agriculture is important, and once the parish was dominated by sugar cane and sugar factories. Negril. in the far west, with its fabulous beaches, is now one of the most popular tourist areas of Jamaica.

ST. ELIZABETH

Capital: Black River
Area: 1,211 sq km
Population: 151,000 (2011 census)
Main towns: Balaclava, Junction, Malvern, Santa Cruz, Treasure Beach
Farming: sugar cane, cassava, coffee, vegetables

Profile: There are mountains in the north and plains in the centre and south. This parish produces the bulk of Jamaica's ground provisions, fruit and vegetables. Black River has a large shrimp and freshwater fishery. Bauxite is mined and processed and alumina exported via Port Kaiser.

MANCHESTER

Capital: Mandeville
Area: 828 sq km
Population: 191,000 (2011 census)
Main towns: Christiana, Mile Gully, Porus
Farming: citrus, coffee, bananas, pimento, ginger, potatoes

Profile: Manchester is mountainous and is the only parish with a capital not located on the coast. It was established in 1814 from parts of St. Elizabeth, Clarendon and Vere. Bauxite mining is an important industry.

CLARENDON

Capital: May Pen
Area: 1,193 sq km
Population: 246,000 (2011 census)
Main towns: Chapelton, Frankfield, Lionel Town
Farming: tobacco, sugar cane, bananas, fish, cattle

Profile:
Southern Clarendon is dominated by sugar cane with Jamaica's largest sugar factory at Moneymusk. The north of the parish is also agricultural, with citrus, tobacco, cattle and vegetables.

ST. ANN

Capital: St. Ann's Bay
Area: 1,210 sq km
Population: 173,000 (2011 census)
Main towns: Browns Town, Ocho Rios, Runaway Bay
Farming: bananas, pimento, sugar, coconuts, coffee, beef and dairy cattle

Profile: St. Ann is the largest parish in area. The underlying limestone rock has been shaped into many caves and sinkholes – one is the Green Grotto in Runaway Bay. Tourism, and related services, are important, and many cruise ships dock at Ocho Rios Bay. Hotels and resorts line the entire coast of St Ann. It's also called Jamaica's Garden Parish and agriculture remains important. The parish holds vast reserves of bauxite.

ST. MARY

Capital: Port Maria
Area: 611 sq km
Population: 114,000 (2011 census)
Main towns: Annotto Bay, Highgate, Oracabessa
Farming: bananas, citrus, pimento, cocoa, coconut

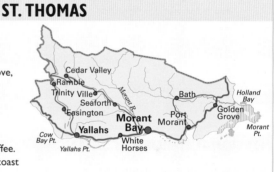

Profile: This was the first part of Jamaica to be settled by the Spanish. After Jamaica was conquered by the English, Puerto Santa Maria became St. Mary. The parish cultivates a wide range of agricultural produce. The parish boasts Jamaica's third international airport named after Ian Fleming who wrote the James Bond books at his home in St Mary. Tourism is becoming increasingly important.

PORTLAND

Capital: Port Antonio
Area: 814 sq km
Population: 82,000 (2011 census)
Main towns: Buff Bay, Orange Bay, Long Bay
Farming: coffee, bananas, sugar cane, citrus, coconut, breadfruit, ackee

Profile: Portland is an area of great natural beauty with stunning beaches, caves, waterfalls and lush vegetation. The Blue Mountain range forms the southern boundary of the parish and is home to communities of Maroons. The coastal strip, with its abundant rainfall, is ideal for agriculture and a wide variety of crops are grown. Jamaica's tourist industry was first established around Port Antonio.

PARISHES

ST. THOMAS

Capital: Morant Bay
Area: 742 sq km
Population: 94,000 (2011 census)
Main towns: Cedar Valley, Golden Grove, Port Morant, Yallahs
Farming: sugar cane, bananas, fruit

Profile: The Blue Mountain range forms the northern boundary, the home of Jamaica's world famous Blue Mountain coffee. Agriculture is also important around the coast and its bananas and sugar are exported.

ST. CATHERINE

Capital: Spanish Town
Area: 1,191 sq km
Population: 518,000 (2011 census)
Main towns: Portmore, Bog Walk, Ewarton, Old Harbour, Linstead
Farming: bananas, coconuts, citrus, pineapples, dairy cattle

Profile: Spanish Town was once the capital of Jamaica (until 1872) and is Jamaica's second largest city. Nearby, is the new city of Portmore, which is the third largest. Both are now closely linked with Kingston. Agriculture continues to be important as St. Catherine has flat fertile lands.

KINGSTON

Capital: Kingston
Area: 23 sq km
Population: 89,000 (2011 census)

Profile: The city of Kingston was established following the destruction of Port Royal in the earthquake and tsunami of 1692. Its natural sheltered harbour, one of the world's finest, is protected by a long sandy spit called The Palisadoes. Jamaica's economic life revolves around Kingston. It is the hub of the financial and service industries and well as being the main cultural centre of the island. The city has spread north and eastward into the neighbouring parish of St. Andrew. As a result, the parishes of Kingston and St. Andrew were amalgamated in 1923 and have been jointly administered since then.

St. Andrew

Capital: Half Way Tree
Area: 435 sq km
Population: 577,000 (2011 census)
Main towns: Half Way Tree, Constant Spring, Harbour View, Stony Hill, Red Hills
Farming: coffee, mango, vegetables, forestry

Profile: St. Andrew is the parish with the highest population. A modern port and industrial complex are located on Kingston Harbour to the west of the old port in downtown Kingston. Gypsum and other minerals are mined in the Bull Bay area in the east. Several universities are in St. Andrew, including the University of the West Indies. Jamaican music, centred on Trench Town, is world famous and a major industry.

POPULATION

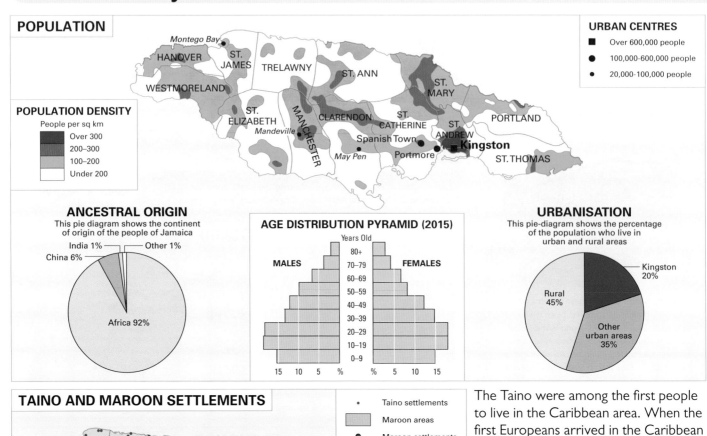

URBAN CENTRES
- ■ Over 600,000 people
- ● 100,000-600,000 people
- • 20,000-100,000 people

POPULATION DENSITY
People per sq km
- Over 300
- 200–300
- 100–200
- Under 200

ANCESTRAL ORIGIN
This pie diagram shows the continent of origin of the people of Jamaica

- India 1%
- Other 1%
- China 6%
- Africa 92%

AGE DISTRIBUTION PYRAMID (2015)

Years Old

MALES · FEMALES

80+
70–79
60–69
50–59
40–49
30–39
20–29
10–19
0–9

15 10 5 % % 5 10 15

URBANISATION
This pie-diagram shows the percentage of the population who live in urban and rural areas

- Kingston 20%
- Rural 45%
- Other urban areas 35%

TAINO AND MAROON SETTLEMENTS

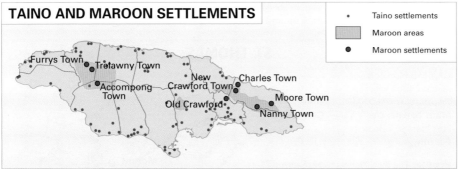

- • Taino settlements
- ▨ Maroon areas
- ● Maroon settlements

Furrys Town, Trelawny Town, Accompong Town, New Crawford Town, Charles Town, Old Crawford, Moore Town, Nanny Town

The Taino were among the first people to live in the Caribbean area. When the first Europeans arrived in the Caribbean in the late 15th century there were over 200 Taino villages in Jamaica. The Maroons are descendants of Africans who escaped from slavery and established their own free communities, mainly in the mountains and in the east of the island.

NATIONAL HEROES OF JAMAICA

The Order of National Hero may be conferred on any person who has rendered heroic service to Jamaica. The first five heroes were proclaimed in 1969. Nanny and Sam Sharpe were proclaimed heroes in 1982. Six of the heroes were born in Jamaica. Nanny was born in Africa and arrived at Port Morant enslaved.

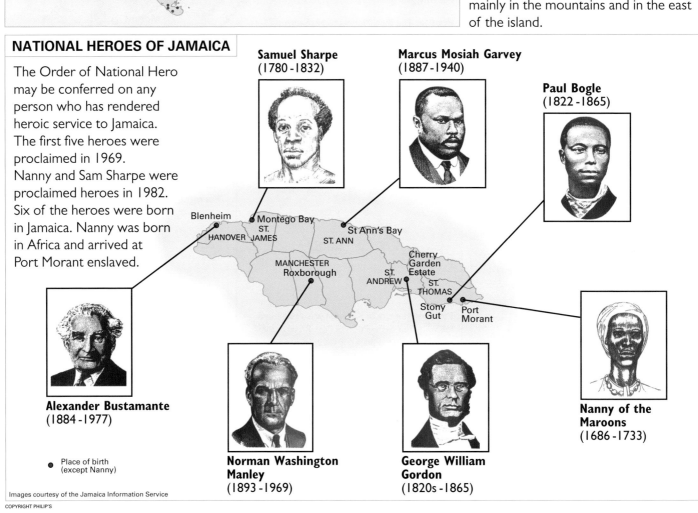

Samuel Sharpe
(1780-1832)

Marcus Mosiah Garvey
(1887-1940)

Paul Bogle
(1822-1865)

Alexander Bustamante
(1884-1977)

Norman Washington Manley
(1893-1969)

George William Gordon
(1820s-1865)

Nanny of the Maroons
(1686-1733)

- ● Place of birth (except Nanny)

CLIMATE

ANNUAL RAINFALL AND PREVAILING WIND

November to April

Montego Bay

Cinchona Gardens

Port Antonio

Kingston

May to October

Montego Bay

Cinchona Gardens

Port Antonio

Kingston

Rainfall in mm

- Over 3000
- 2000–3000
- 1000–2000
- 500–1000
- Under 500
- Prevailing winds (January)
- Prevailing winds (July)

CLIMATE GRAPHS

Name of place — **KINGSTON**

Average monthly temperature

Average annual rainfall — Rainfall 688mm

Average monthly rainfall

Months of the year — J F M A M J J A S O N D

MONTEGO BAY — Rainfall 831mm

CINCHONA GDNS — Rainfall 1500mm

PORT ANTONIO — Rainfall 2335mm

AGRICULTURE

-)) Bananas
- ◖ Coconuts
- ⌁ Sugar

- Forest
- Scrub and livestock
- Agriculture
- Wetland
- Urban area

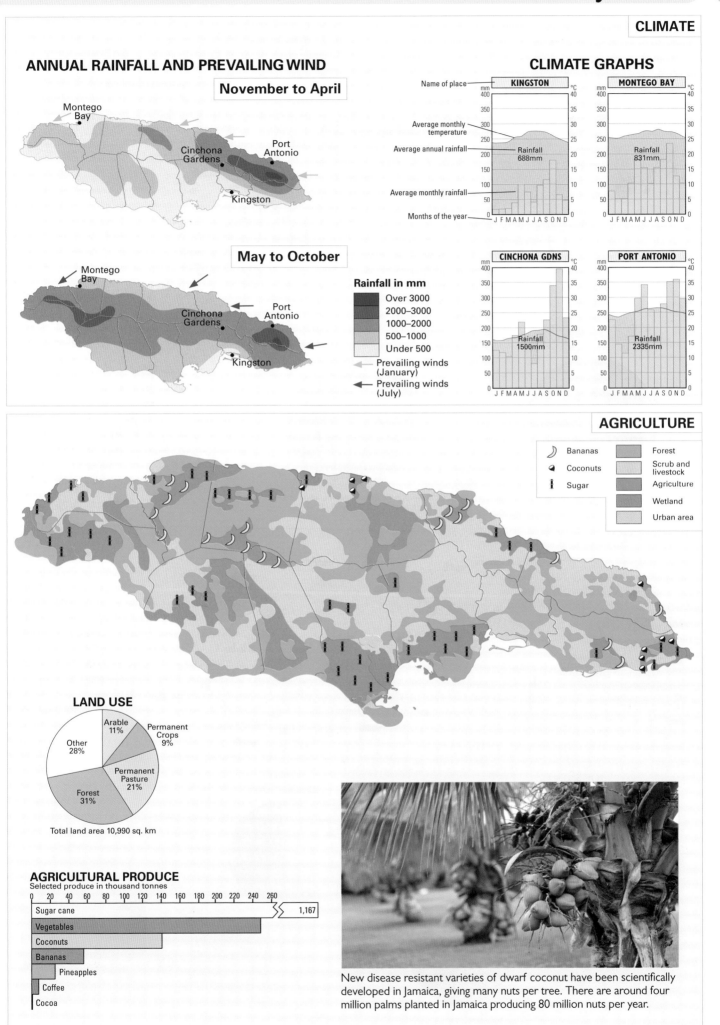

LAND USE

- Arable 11%
- Permanent Crops 9%
- Permanent Pasture 21%
- Forest 31%
- Other 28%

Total land area 10,990 sq. km

AGRICULTURAL PRODUCE

Selected produce in thousand tonnes

0 20 40 60 80 100 120 140 160 180 200 220 240 260

- Sugar cane — 1,167
- Vegetables
- Coconuts
- Bananas
- Pineapples
- Coffee
- Cocoa

New disease resistant varieties of dwarf coconut have been scientifically developed in Jamaica, giving many nuts per tree. There are around four million palms planted in Jamaica producing 80 million nuts per year.

TOURISM

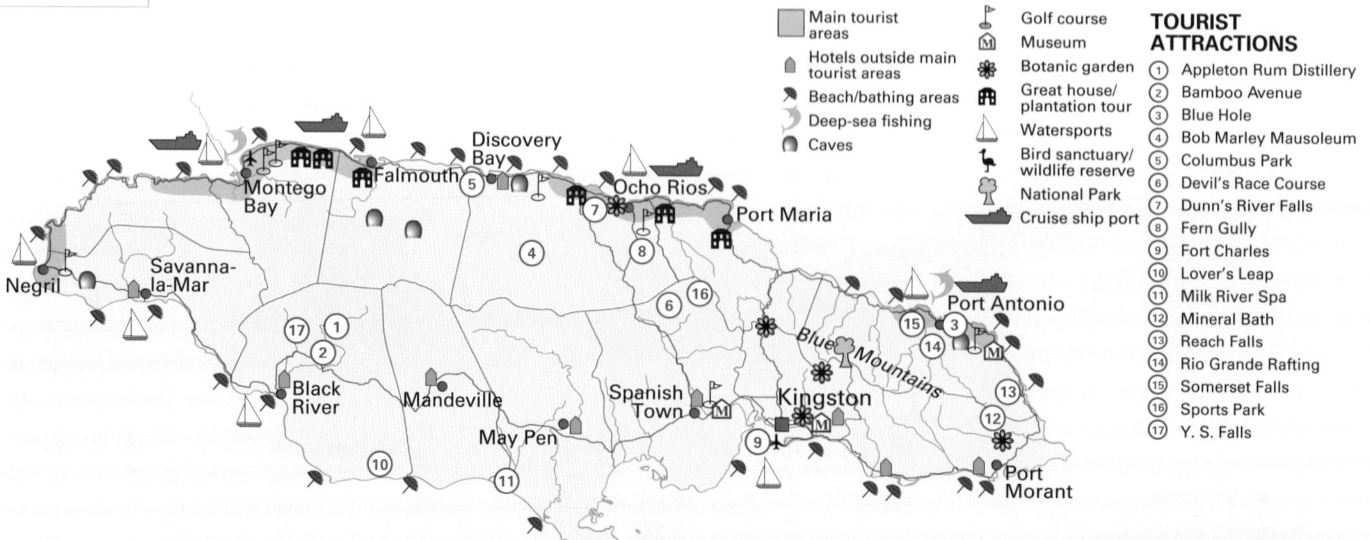
Main tourist areas
Hotels outside main tourist areas
Beach/bathing areas
Deep-sea fishing
Caves
Golf course
Museum
Botanic garden
Great house/ plantation tour
Watersports
Bird sanctuary/ wildlife reserve
National Park
Cruise ship port
TOURIST ATTRACTIONS
1 Appleton Rum Distillery
2 Bamboo Avenue
3 Blue Hole
4 Bob Marley Mausoleum
5 Columbus Park
6 Devil's Race Course
7 Dunn's River Falls
8 Fern Gully
9 Fort Charles
10 Lover's Leap
11 Milk River Spa
12 Mineral Bath
13 Reach Falls
14 Rio Grande Rafting
15 Somerset Falls
16 Sports Park
17 Y. S. Falls
Montego Bay
Falmouth
Discovery Bay
Ocho Rios
Port Maria
Negril
Savanna-la-Mar
Black River
Mandeville
May Pen
Spanish Town
Kingston
Blue Mountains
Port Antonio
Port Morant

STOPOVER AND CRUISE PASSENGER ARRIVALS

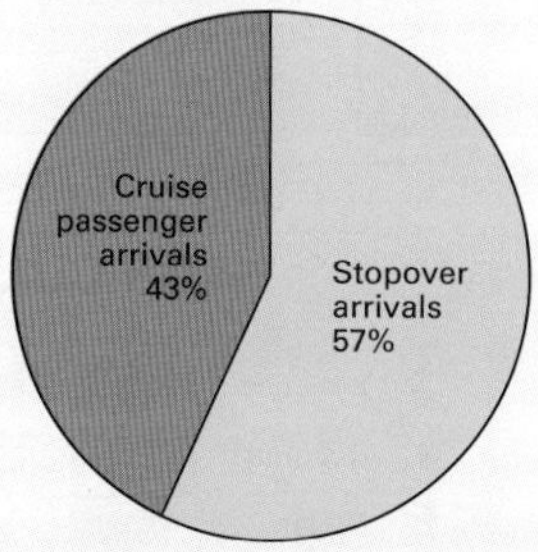
Cruise passenger arrivals 43%
Stopover arrivals 57%

Total visitor arrivals in 2015 3,837,243

ORIGIN OF STOPOVER TOURISTS

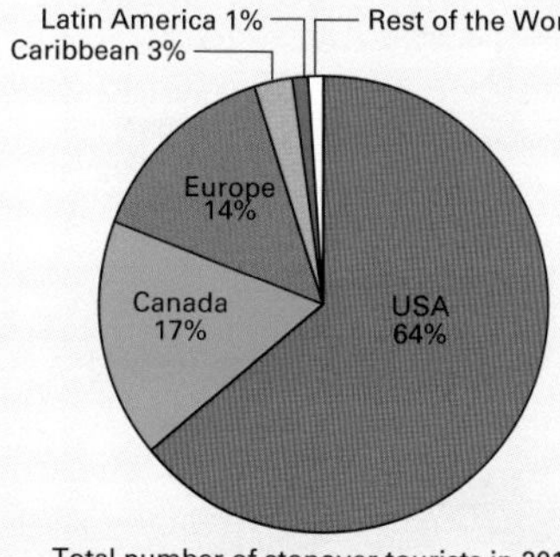
Latin America 1%
Rest of the World 1%
Caribbean 3%
Europe 14%
Canada 17%
USA 64%

Total number of stopover tourists in 2016 2,181,700

RESORT AREA FOR STOPOVER TOURISTS

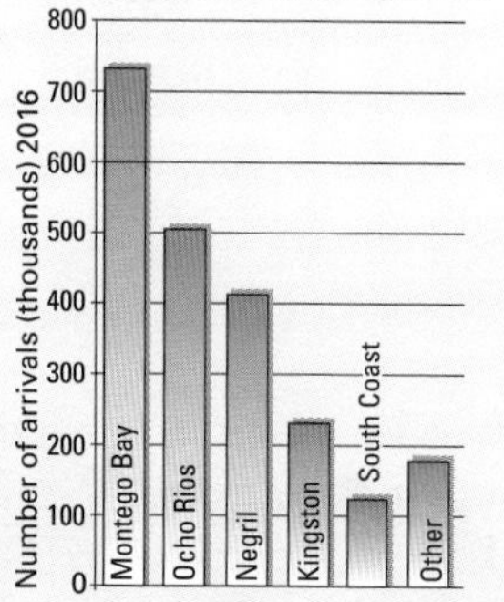
Number of arrivals (thousands) 2016
800
700
600
500
400
300
200
100
0
Montego Bay
Ocho Rios
Negril
Kingston
South Coast
Other

RESOURCES

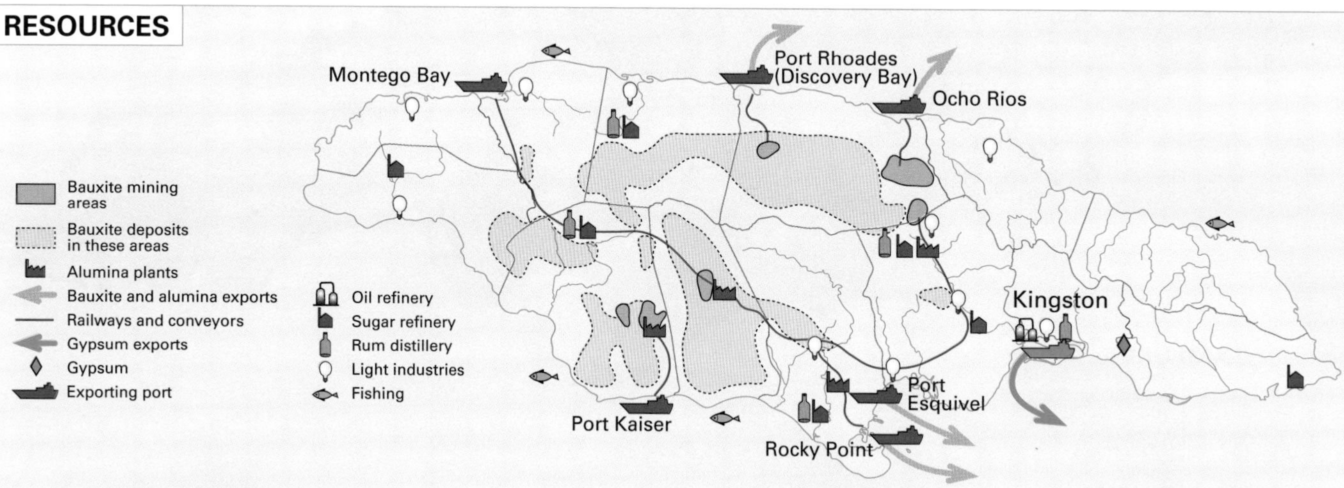
Bauxite mining areas
Bauxite deposits in these areas
Alumina plants
Bauxite and alumina exports
Railways and conveyors
Gypsum exports
Gypsum
Exporting port
Oil refinery
Sugar refinery
Rum distillery
Light industries
Fishing
Montego Bay
Port Rhoades (Discovery Bay)
Ocho Rios
Kingston
Port Esquivel
Port Kaiser
Rocky Point

TOP 5 EXPORTS DESTINATIONS 2016

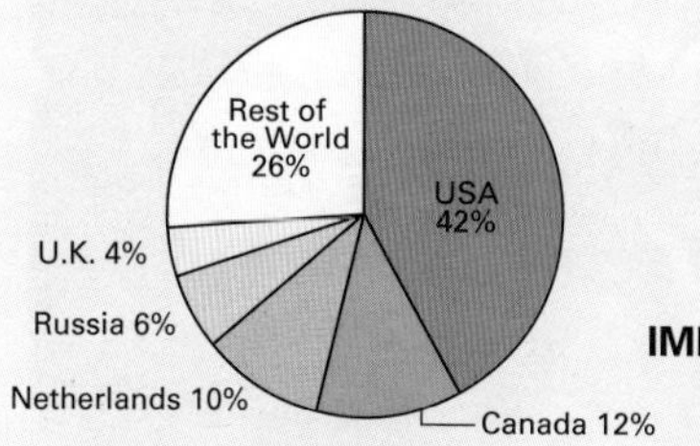
Rest of the World 26%
USA 42%
U.K. 4%
Russia 6%
Netherlands 10%
Canada 12%

Total US$ 1,200million

TOP 5 IMPORT ORIGINS 2016

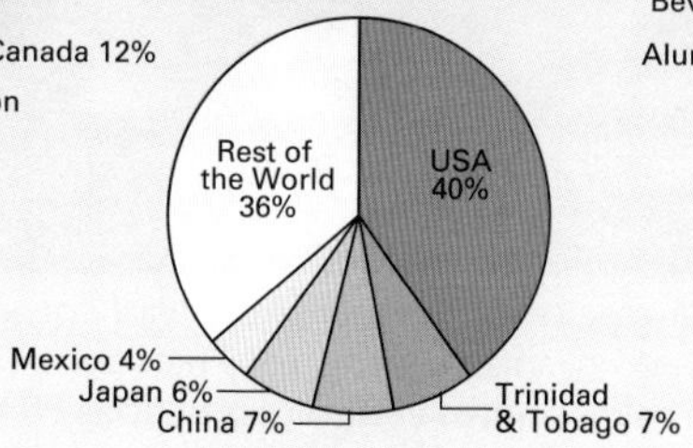
Rest of the World 36%
USA 40%
Mexico 4%
Japan 6%
China 7%
Trinidad & Tobago 7%

Total US$ 4,740million

TOP 5 EXPORTS 2016

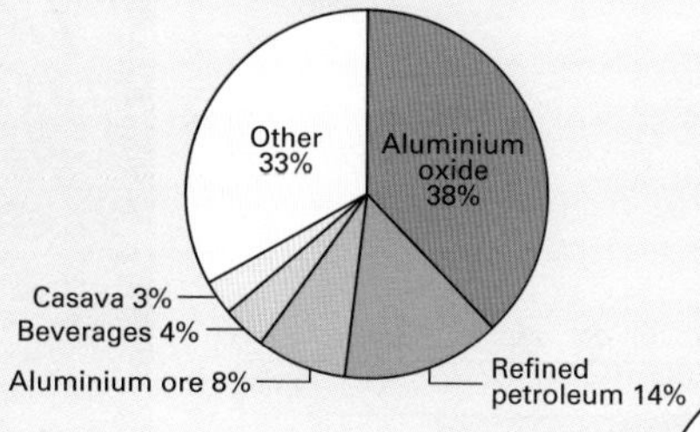
Other 33%
Aluminium oxide 38%
Casava 3%
Beverages 4%
Aluminium ore 8%
Refined petroleum 14%

TOP 5 IMPORTS 2016

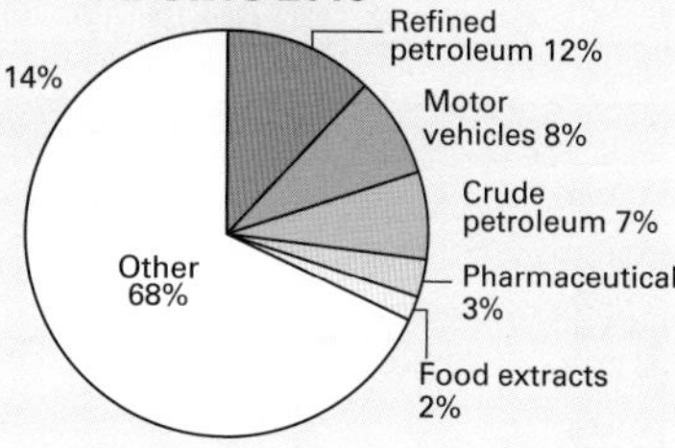
Refined petroleum 12%
Motor vehicles 8%
Crude petroleum 7%
Pharmaceuticals 3%
Food extracts 2%
Other 68%

The combination of plants and animals in a particular area is known as an ecosystem. Land-based ecosystems include forests, deserts, grasslands and mountains. Lakes, wetlands and coral reefs are all water-based ecosystems. Biodiversity is the name given to the variety of plants and animals in an ecosystem.

Jamaica has a rich and varied biodiversity and a number of areas are so important that they have been internationally recognized as Biodiversity Hotspots. These are shown on the map below.

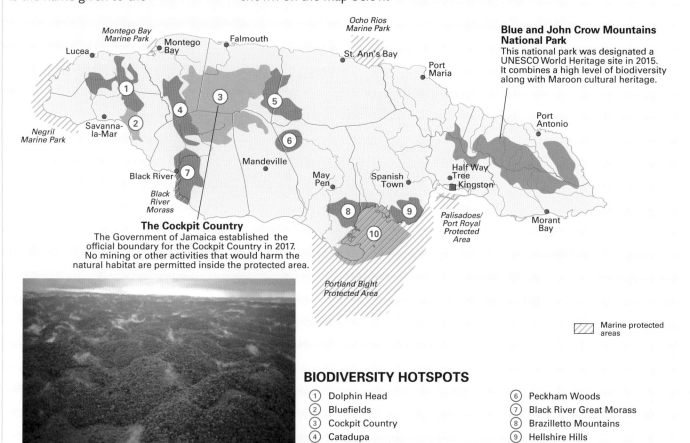

Blue and John Crow Mountains National Park
This national park was designated a UNESCO World Heritage site in 2015. It combines a high level of biodiversity along with Maroon cultural heritage.

The Cockpit Country
The Government of Jamaica established the official boundary for the Cockpit Country in 2017. No mining or other activities that would harm the natural habitat are permitted inside the protected area.

Marine protected areas

Cockpit Country is an area of eroded limestone.

BIODIVERSITY HOTSPOTS

① Dolphin Head		⑥ Peckham Woods	
② Bluefields		⑦ Black River Great Morass	
③ Cockpit Country		⑧ Brazilletto Mountains	
④ Catadupa		⑨ Hellshire Hills	
⑤ Litchfield Mountain - Matheson's Run		⑩ Portland Ridge and Bight	

Jamaica's rich and unique culture includes its language and literature, its creative arts, its sporting accomplishments and its religious practices. This is the result of the diverse origin of Jamaicans. Some aspects of Jamaica's culture are noted below. What other aspects of Jamaican culture do you know?

MUSIC: Jamaican mento music of the 1940s evolved to ska, reggae and dancehall. Reggae as a form of music caught the interest of the world in the 1970s. Bob Marley, pictured here in 1978, is an international icon.

ARTS: Jamaicans excel in creative arts such as dance, drama and other visual arts. This is the National Dance Theatre company in performance. Jamaican film and theatre use the national language, patois, as well as the English-language.

FOOD: Jamaica's unique dishes are a result of the mix of cultures – Tiano, African, European, Indian and Chinese. This dish includes ackee which was introduced to Jamaica by Africans when they were transported in slave ships.

SPORT: Jamaica has excelled in track and field internationally since the 1940s. Usain Bolt, from rural Trelawny, is considered the world's fastest man. In which other sports have Jamaicans gained international acclaim?

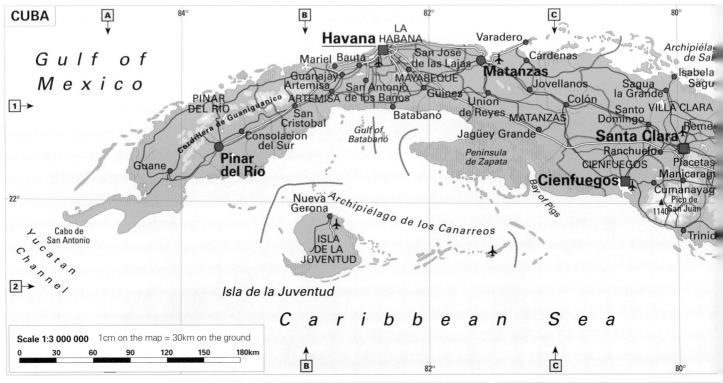

CUBA

Gulf of Mexico

84° | 82° | 80°

Havana LA HABANA

Varadero

Mariel Bauta San José de las Lajas · Cárdenas · Archipiélago de Sal

Guanajay Artemisa MAYABEQUE **Matanzas** Jovellanos Isabela Sagu

PINAR DEL RÍO ARTEMISA San Antonio de los Baños Güines Colón Sagua la Grande Santo VILLA CLARA Reme

San Cristóbal Batabanó Union de Reyes MATANZAS Domingo **Santa Clara**

Consolacion del Sur Jagüey Grande Ranchuelo CIENFUEGOS Placetas Manicarag

Pinar del Río Peninsula de Zapata Bay of Pigs **Cienfuegos** Cumanayag Pico de 1140 San Juan

Guane Trinid

Cordillera de Guaniguanico

Cabo de San Antonio

Yucatan Channel

Nueva Gerona *Archipiélago de los Canarreos*

ISLA DE LA JUVENTUD

Gulf of Batabanó

Isla de la Juventud

C a r i b b e a n   S e a

Scale 1:3 000 000 1cm on the map = 30km on the ground

0   30   60   90   120   150   180km

CUBA

Independence 1902
Capital: Havana
Area: 110,860 sq km
Population: 11,239,000 (est. 2018)
Languages: Spanish
Sources of national income: sugar, nickel, tobacco, fish, petroleum, citrus, coffee

HAITI

Independence 1804
Capital: Port-au-Prince
Area: 27,275 sq km
Population: 10,912,000 (est. 2018)
Languages: French (official), Creole (official)
Sources of national income: clothing, coffee, oils, cocoa

DOMINICAN REPUBLIC

Independence 1844
Capital: Santo Domingo
Area: 48,730 sq km
Population: 10,169,000 (est. 2018)
Languages: Spanish
Sources of national income: ferronickel, su gold, silver, coffee, cocoa, tobacco, meats, consur goods

---

HAITI & DOMINICAN REPUBLIC

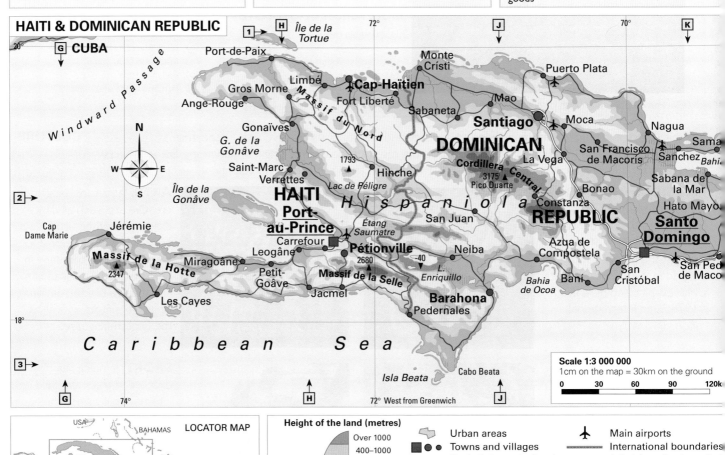

20° | 72° | 70°

CUBA

Île de la Tortue

Port-de-Paix Monte Crísti Puerto Plata

Limbé **Cap-Haïtien** Mao

Gros Morne Fort Liberté Sabaneta **Santiago** Moca Nagua

Ange-Rouge Sabaneta **DOMINICAN** San Francisco Sama de Macorís Sanchez Bahi

Massif du Nord La Vega

Gonaïves *Cordillera Central*

G. de la Gonâve 1793 3175 Pico Duarte Bonao Sabana de la Mar

Saint-Marc Hinche La Vega Constanza **REPUBLIC** Hato Mayo

Verrettes *Lac de Péligre* *H i s p a n i o l a* **Santo Domingo**

Île de la Gonâve **HAITI** San Juan

Port-au-Prince *Étang Saumatre* Azua de Compostela San Ped

Carrefour Pétionville Neiba Baní de Maco

Cap Dame Marie Jérémie Leogâne 2680 -40 San Cristóbal

Massif de la Hotte Miragoâne *Massif de la Selle* L. Enriquillo Bahia de Ocoa

2347 Petit-Goâve **Barahona** Isla Beata

Les Cayes Jacmel Pedernales Cabo Beata

Windward Passage

N W E S

C a r i b b e a n   S e a

74° | 72° West from Greenwich

Scale 1:3 000 000 1cm on the map = 30km on the ground

0   30   60   90   120k

LOCATOR MAP

USA BAHAMAS
MEXICO CUBA
DOMINICAN REPUBLIC
BELIZE HAITI PUERTO RICO & Virgin Is.
JAMAICA Hispaniola

COPYRIGHT PHILIP'S

Height of the land (metres)

Over 1000
400–1000
200–400
100–200
0–100
Sea level
Below sea level

Urban areas
Towns and villages
Havana Capital city underlined
Highways
Roads
Railways

Main airports
International boundaries
Administrative boundari
Mangroves
Reefs

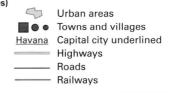

PUERTO RICO (U.S.A.)

S.A. Unincorporated Territory
apital: San Juan
rea: 9,104 sq km
opulation: 3,678,000 (est.2018)
nguages: Spanish, English
urces of national income: sugar cane,
ffee, fruit, food products, chemicals, electronics, clothing, tourism

UNITED STATES VIRGIN ISLANDS

U.S.A. Unincorporated Territory
Capital: Charlotte Amalie
Area: 352 sq km
Population: 106,000 (est. 2018)
Languages: English (official), Spanish, Creole
Sources of national income: refined petroleum
products

PUERTO RICO & US VIRGIN ISLANDS

Scale 1:2 000 000
1cm on the map = 20km on the ground

0   20   40   60   80km

COPYRIGHT PHILIP'S

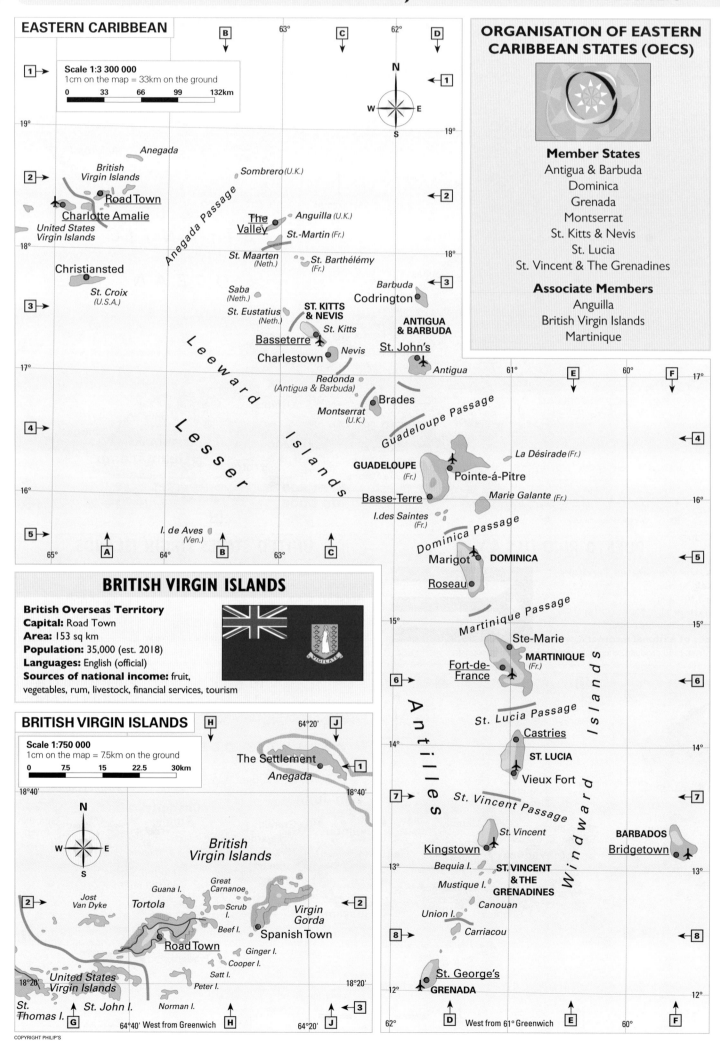

EASTERN CARIBBEAN

Scale 1:3 300 000
1cm on the map = 33km on the ground

0   33   66   99   132km

Anegada

British
Virgin Islands

Sombrero (U.K.)

Road Town
Charlotte Amalie

United States
Virgin Islands

The
Valley

Anguilla (U.K.)

St.-Martin (Fr.)

St. Maarten
(Neth.)

St. Barthélémy
(Fr.)

Christiansted

St. Croix
(U.S.A.)

Saba
(Neth.)

St. Eustatius
(Neth.)

Barbuda

Codrington

**ST. KITTS
& NEVIS**

St. Kitts

**ANTIGUA
& BARBUDA**

Basseterre

Nevis

St. John's

Charlestown

Antigua

Redonda
(Antigua & Barbuda)

Brades

Montserrat
(U.K.)

GUADELOUPE
(Fr.)

La Désirade (Fr.)

Pointe-á-Pitre

Basse-Terre

Marie Galante (Fr.)

I. des Saintes
(Fr.)

I. de Aves
(Ven.)

Dominica Passage

Marigot   **DOMINICA**

Roseau

Anegada Passage

Leeward

Lesser

Islands

Guadeloupe Passage

Martinique Passage

Ste-Marie

Fort-de-
France

MARTINIQUE
(Fr.)

St. Lucia Passage

Castries

ST. LUCIA

Vieux Fort

St. Vincent Passage

St. Vincent

Kingstown

Bequia I.

**ST. VINCENT
& THE
GRENADINES**

Mustique I.

Canouan

Union I.

Carriacou

St. George's

GRENADA

BARBADOS
Bridgetown

Antilles

Windward   Islands

West from 61° Greenwich

ORGANISATION OF EASTERN CARIBBEAN STATES (OECS)

Member States
Antigua & Barbuda
Dominica
Grenada
Montserrat
St. Kitts & Nevis
St. Lucia
St. Vincent & The Grenadines

Associate Members
Anguilla
British Virgin Islands
Martinique

BRITISH VIRGIN ISLANDS

British Overseas Territory
Capital: Road Town
Area: 153 sq km
Population: 35,000 (est. 2018)
Languages: English (official)
Sources of national income: fruit,
vegetables, rum, livestock, financial services, tourism

BRITISH VIRGIN ISLANDS

Scale 1:750 000
1cm on the map = 7.5km on the ground

0   7.5   15   22.5   30km

The Settlement

Anegada

British
Virgin Islands

Guana I.

Great
Carnanoe

Jost
Van Dyke

Tortola

Scrub
I.

Beef I.

Virgin
Gorda

Spanish Town

Road Town

Ginger I.

Cooper I.

Satt I.

Peter I.

United States
Virgin Islands

St.
Thomas I.

St. John I.

Norman I.

64°40' West from Greenwich   64°20'

ANGUILLA

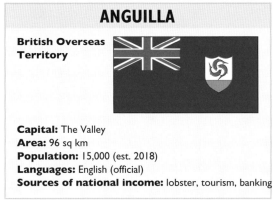

British Overseas Territory

Capital: The Valley
Area: 96 sq km
Population: 15,000 (est. 2018)
Languages: English (official)
Sources of national income: lobster, tourism, banking

Sandy Ground, on the west coast, is one of the centres of tourism in Anguilla.

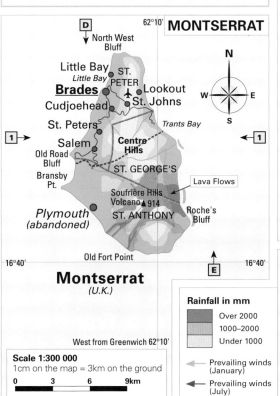

ANGUILLA

Scale 1:300 000
1cm on the map = 3km on the ground

0   3   6   9   12km

Prickly Pear Cays
Seal I.
Shoal Bay Village
Island Harbour
Snake Pt.
Scrub I.
Anguilla (U.K.)
Flat Cap Pt.
North Side 59 ▲
Savannah Bay
Gibbon Pt.
East End Village
Crocus Bay
The Valley
Sandy Hill Bay
Sandy I.
George Hill
Sandy Hill Bay
Sandy Ground Village
The Quarter
Mead Pt.
Forest Pt.
Long Bay Village
South Hill Village
West End Village
Blowing Point Village
Rendezvous Bay
Shaddick Pt.
Blowing Rock
18°10'
18°10'
63°10'
West from Greenwich 63°

RESOURCES

- ☂ Tourism
- ▨ Forest
- ▢ Golf course
- ▨ Scrub and livestock
- ▨ Urban area

Sandy Ground

MONTSERRAT

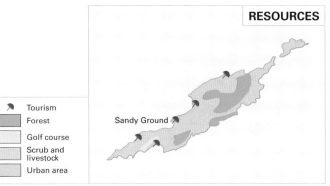

British Overseas Territory
Capital: Brades
Area: 102 sq km
Population: 5,000 (est. 2018)
Languages: English

MONTSERRAT

62°10'
North West Bluff
Little Bay
Little Bay
ST. PETER
Brades
Lookout
Cudjoehead
St. Johns
St. Peters
Trants Bay
Salem
Centre Hills
Old Road Bluff
ST. GEORGE'S
Bransby Pt.
Soufrière Hills Volcano ▲914
Roche's Bluff
Plymouth (abandoned)
ST. ANTHONY
Old Fort Point
16°40'
16°40'

Montserrat (U.K.)

West from Greenwich 62°10'

Scale 1:300 000
1cm on the map = 3km on the ground
0   3   6   9km

VOLCANIC ERUPTIONS

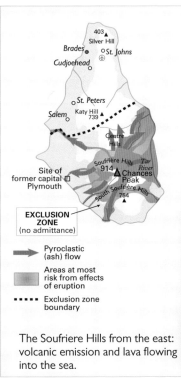

403 ▲ Silver Hill
Brades
○ St. Johns
Cudjoehead
⊕
○ St. Peters
Salem Katy Hill 739
Centre Hills
Soufrière Hills 914 ▲ Chances Peak
Tar River
Site of former capital ☐ Plymouth
South Soufrière Hills 754 ▲

EXCLUSION ZONE (no admittance)

⟹ Pyroclastic (ash) flow
▨ Areas at most risk from effects of eruption
▪▪▪▪ Exclusion zone boundary

In 1995, after almost 400 years lying dormant, the Soufrière Hills volcano began a series of eruptions. Further eruptions in 1996 and 1997 left the south of the island uninhabitable and 5,000 people had to be evacuated to the northern zone.

The Soufriere Hills from the east: volcanic emission and lava flowing into the sea.

RAINFALL

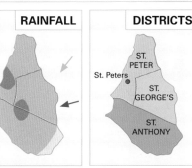

Rainfall in mm
- ▨ Over 2000
- ▨ 1000–2000
- ▢ Under 1000

⟶ Prevailing winds (January)
⟶ Prevailing winds (July)

DISTRICTS

ST. PETER
St. Peters ●
ST. GEORGE'S
ST. ANTHONY

RESOURCES

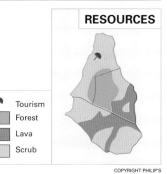

- ☂ Tourism
- ▨ Forest
- ▨ Lava
- ▢ Scrub

Lava Flows

COPYRIGHT PHILIP'S

ANTIGUA & BARBUDA

Independence 1981
Capital: St John's
Area: 442 sq km
Population: 102,000 (est. 2018)
Languages: English (official), local dialects
Sources of national income: local agriculture, tourism, manufactured goods and handicrafts

PARISHES

St. John's
ST. JOHN
ST. GEORGE
Pigotts
Parham
ST. PETER
Bolans
ST. MARY
ST. PAUL
ST. PHILIP
Falmouth

ANTIGUA

Antigua

Beggars Pt.
Cedar Grove
17°10'
Fort James
St. Johnston Village
Pigotts
ST. GEORGE
Parham
Parham Harb.
Long I.
Great Bird I.
Guiana I.
Crump I.
Pelican I.
Indian Town Pt.
St. John's
ST. JOHN
Potters Village
Pares
Seatons
Willikies
Five Islands Village
Five I. Harbour
Jennings
Bolans
Buckleys
Swetes
All Saints
ST. PETER
ST. PHILIP
Nonsuch Bay
Green I.
Newfield
Freetown
York I.
Mount Obama ▲ 402
ST. MARY
Liberta
▲ 368
ST. PAUL
Bethesda
Johnsons Point
Urlings
Falmouth
Soldier Pt.
Johnsons Pt.
Old Road
English Harbour Town
Falmouth Harbour
Willoughby Bay
17°
Old Road Bluff
Nanton Pt.

Guadeloupe Passage

West from Greenwich 61°40'

Scale 1:250 000
1cm on the map = 2.5km on the ground
0   2.5   5   7.5   10km

Legend (agriculture/resources symbols)
- Bananas
- Coconuts
- Pineapples
- Rum distillery
- Oil refinery
- Tourism
- Major port
- Forest
- Scrub and livestock
- Agriculture
- Urban area

Rainfall in mm
- Over 1000
- Under 1000
- Prevailing winds (January)
- Prevailing winds (July)

RAINFALL

RESOURCES

LOCATOR MAP
Anguilla
St. Maarten
Saba
St. Eustatius
Barbuda
ANTIGUA & BARBUDA
Antigua
ST. KITTS AND NEVIS
Caribbean Sea
Montserrat
GUADELOUPE

BARBUDA

Scale 1:300 000
1cm on the map = 3km on the ground
0   3   6   9   12km

Goat Pt.
Billy Pt.
Goat I.
Kid I.
Hog Pt.
61°50'
Cedar Tree Pt.
Codrington Lagoon
17°40'
The Highlands ▲ 39
Codrington
Barbuda
Dulcina
Palmetto Pt.
Cocoa Point
Gravenor Bay
Spanish Pt.
West from 61°50' Greenwich
61°40'

RESOURCES

- Tourism
- Scrub
- Urban area

RAINFALL

Rainfall in mm
- Under 1500
- Prevailing winds (January)
- Prevailing winds (July)

Height of the land (metres)
- Over 1000
- 400–1000
- 200–400
- 100–200
- 0–100
- Below sea level
- Sea level

- Towns and villages
- St. John's Capital city underlined
- Roads
- Main airports
- Administrative boundaries
- Mangroves
- Reefs

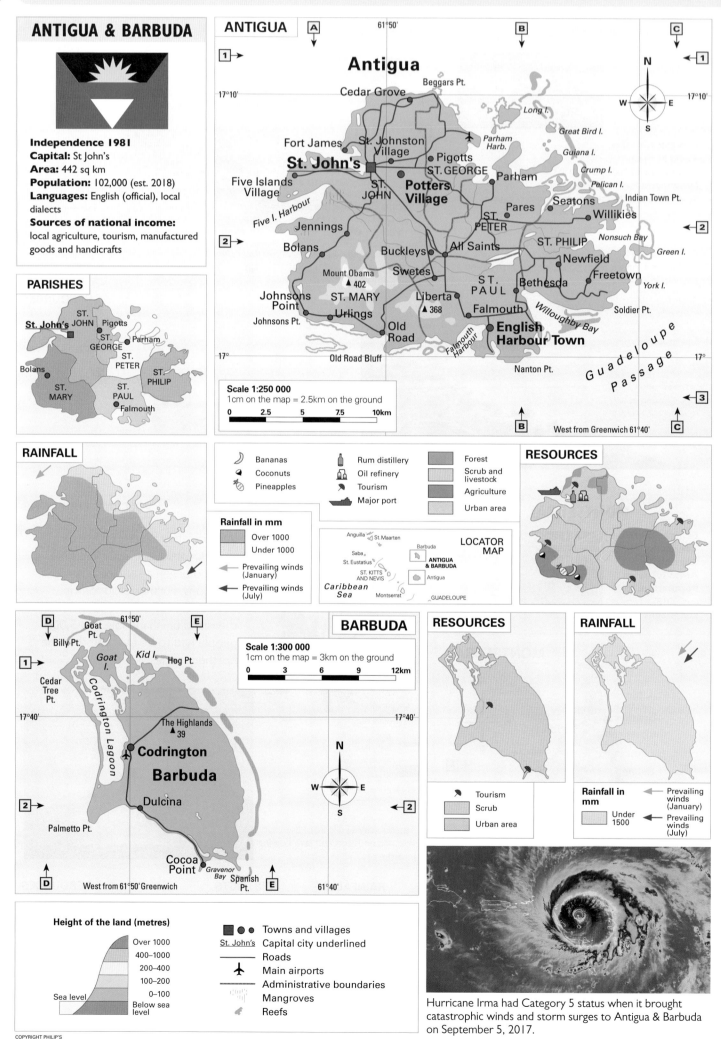

Hurricane Irma had Category 5 status when it brought catastrophic winds and storm surges to Antigua & Barbuda on September 5, 2017.

ST. KITTS & NEVIS

ST. KITTS & NEVIS

Independence 1983
Capital: Basseterre
Area: 261 sq km
Population: 55,000 (est. 2018)
Languages: English
Sources of national income:
bananas, fish, tourism, offshore
finance, forestry

Height of the land (metres)

Over 1000
400–1000
200–400
100–200
Sea level   0–100
Below sea
level

Urban areas
Towns and villages
Basseterre Capital city underlined
Highways
Roads
Railways
Main airports
Administrative
boundaries
Mangroves
Reefs

Looking out to sea from the
historic Brimstone Hill fortress.

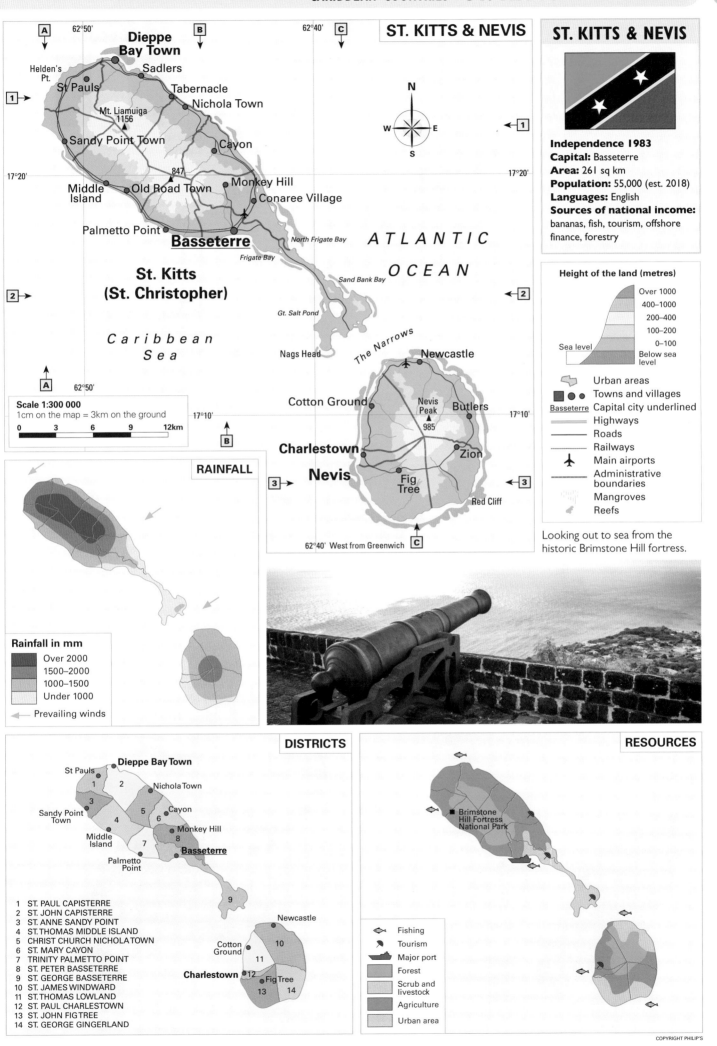

Map — St. Kitts & Nevis

Dieppe Bay Town
Helden's Pt.
Sadlers
St Pauls
Tabernacle
Nichola Town
Mt. Liamuiga 1156
Sandy Point Town
Cayon
847
Middle Island
Old Road Town
Monkey Hill
Conaree Village
Palmetto Point
Basseterre
North Frigate Bay

St. Kitts (St. Christopher)
Frigate Bay
Sand Bank Bay

A T L A N T I C   O C E A N

Caribbean Sea

Gt. Salt Pond

Nags Head
The Narrows
Newcastle
Cotton Ground
Nevis Peak 985
Butlers
Charlestown
Zion
Nevis
Fig Tree
Red Cliff

62°40' West from Greenwich

Scale

Scale 1:300 000
1cm on the map = 3km on the ground
0   3   6   9   12km

RAINFALL

Rainfall in mm
Over 2000
1500–2000
1000–1500
Under 1000
Prevailing winds

DISTRICTS

St Pauls
Dieppe Bay Town
Nichola Town
Cayon
Sandy Point Town
Middle Island
Monkey Hill
Basseterre
Palmetto Point

Newcastle
Cotton Ground
Charlestown
Fig Tree

1 ST. PAUL CAPISTERRE
2 ST. JOHN CAPISTERRE
3 ST. ANNE SANDY POINT
4 ST. THOMAS MIDDLE ISLAND
5 CHRIST CHURCH NICHOLA TOWN
6 ST. MARY CAYON
7 TRINITY PALMETTO POINT
8 ST. PETER BASSETERRE
9 ST. GEORGE BASSETERRE
10 ST. JAMES WINDWARD
11 ST. THOMAS LOWLAND
12 ST. PAUL CHARLESTOWN
13 ST. JOHN FIG TREE
14 ST. GEORGE GINGERLAND

RESOURCES

Brimstone
Hill Fortress
National Park

Fishing
Tourism
Major port
Forest
Scrub and livestock
Agriculture
Urban area

COPYRIGHT PHILIP'S

GUADELOUPE

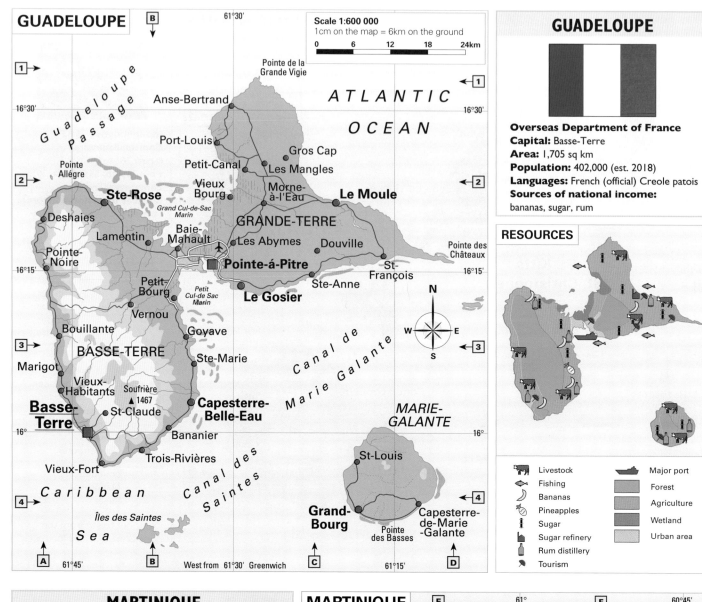

GUADELOUPE

Overseas Department of France
Capital: Basse-Terre
Area: 1,705 sq km
Population: 402,000 (est. 2018)
Languages: French (official) Creole patois
Sources of national income:
bananas, sugar, rum

RESOURCES

Livestock
Fishing
Bananas
Pineapples
Sugar
Sugar refinery
Rum distillery
Tourism
Major port
Forest
Agriculture
Wetland
Urban area

Scale 1:600 000
1cm on the map = 6km on the ground
0   6   12   18   24km

MARTINIQUE

MARTINIQUE

Overseas Department of France
Capital: Fort-de-France
Area: 1,102 sq km
Population: 386,000 (est. 2018)
Languages: French (official) Creole patois
Sources of national income: petroleum products, rum

RESOURCES

Livestock
Fishing
Bananas
Pineapples
Sugar
Sugar refinery
Rum distillery
Oil refinery
Tourism
Major port

Forest
Agriculture
Wetland
Urban area

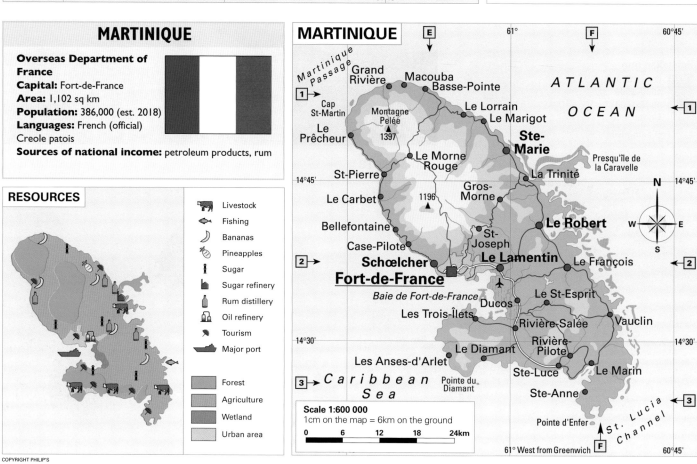

Scale 1:600 000
1cm on the map = 6km on the ground
0   6   12   18   24km

DOMINICA

Independence 1978
Capital: Roseau
Area: 751 sq km
Population: 74,000 (est. 2018)
Languages: English (official) French patois
Sources of national income: bananas, citrus, coconut oil, soap, ecotourism

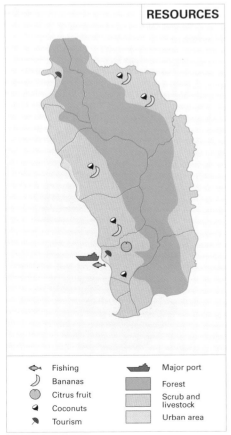

Dominica farmers produce food to export to neighbouring islands such as dasheen, yam, tania, plantain and citrus.

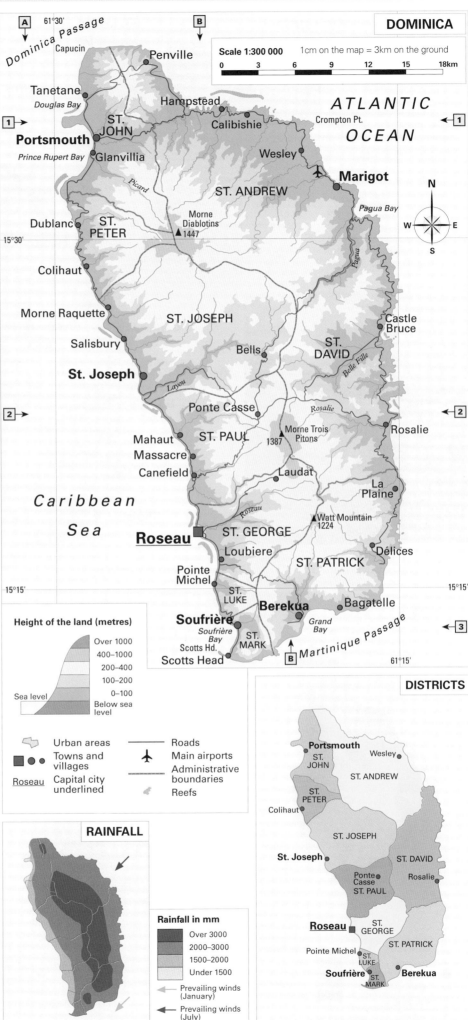

DOMINICA

Scale 1:300 000   1cm on the map = 3km on the ground

| 0 | 3 | 6 | 9 | 12 | 15 | 18km |

61°30'

Dominica Passage

Capucin

Penville

Tanetane

Douglas Bay

Hampstead

ST. JOHN

Calibishie

ATLANTIC OCEAN

Crompton Pt.

Portsmouth

Prince Rupert Bay

Glanvillia

Wesley

Marigot

Picard

Dublanc

ST. PETER

ST. ANDREW

Morne Diablotins ▲ 1447

Pagua Bay

15°30'

Colihaut

Pagua

Morne Raquette

ST. JOSEPH

Castle Bruce

Salisbury

Bells

ST. DAVID

Belle Fille

St. Joseph

Layou

Rosalie

Ponte Casse

ST. PAUL

Morne Trois Pitons 1387

Rosalie

Mahaut

Massacre

Canefield

Laudat

La Plaine

Caribbean Sea

Roseau

Watt Mountain 1224

Roseau

ST. GEORGE

Loubiere

Délices

Pointe Michel

ST. PATRICK

ST. LUKE

Berekua

Bagatelle

15°15'

Soufrière

Soufrière Bay

ST. MARK

Grand Bay

Scotts Hd.

Scotts Head

Martinique Passage

61°15'

N W E S

RESOURCES

Height of the land (metres)

- Over 1000
- 400–1000
- 200–400
- 100–200
- 0–100
- Below sea level

Sea level

Urban areas
Towns and villages
Roseau   Capital city underlined

Roads
Main airports
Administrative boundaries
Reefs

- Fishing
- Bananas
- Citrus fruit
- Coconuts
- Tourism

- Major port
- Forest
- Scrub and livestock
- Urban area

RAINFALL

Rainfall in mm

- Over 3000
- 2000–3000
- 1500–2000
- Under 1500

→ Prevailing winds (January)
→ Prevailing winds (July)

DISTRICTS

Portsmouth
ST. JOHN
Wesley
ST. ANDREW
ST. PETER
Colihaut
ST. JOSEPH
St. Joseph
ST. DAVID
Ponte Casse
ST. PAUL
Rosalie
Roseau ST. GEORGE
ST. PATRICK
Pointe Michel
ST. LUKE
Soufrière ST. MARK
Berekua

COPYRIGHT PHILIP'S

ST. LUCIA

Independence 1979
Capital: Castries
Area: 616 sq km
Population: 180,000
(est. 2018)
Languages: English (official),
French patois
Sources of national income: bananas, coconuts, mangos,
avocados, tourism, manufacturing

Height of the land (metres)

Over 1000
400–1000
200–400
100–200
0–100
Sea level
Below sea level

Towns and villages
Castries — Capital city underlined
Roads
Main airports
Administrative boundaries
Reefs

About 400 cruise ships visit St. Lucia each year
bringing 700,000 passengers to visit.

RESOURCES

Bananas
Coconuts
Cocoa
Oil refinery
Rum distillery
Tourism

Major port
Forest
Scrub and livestock
Agriculture
Urban area

COPYRIGHT PHILIP'S

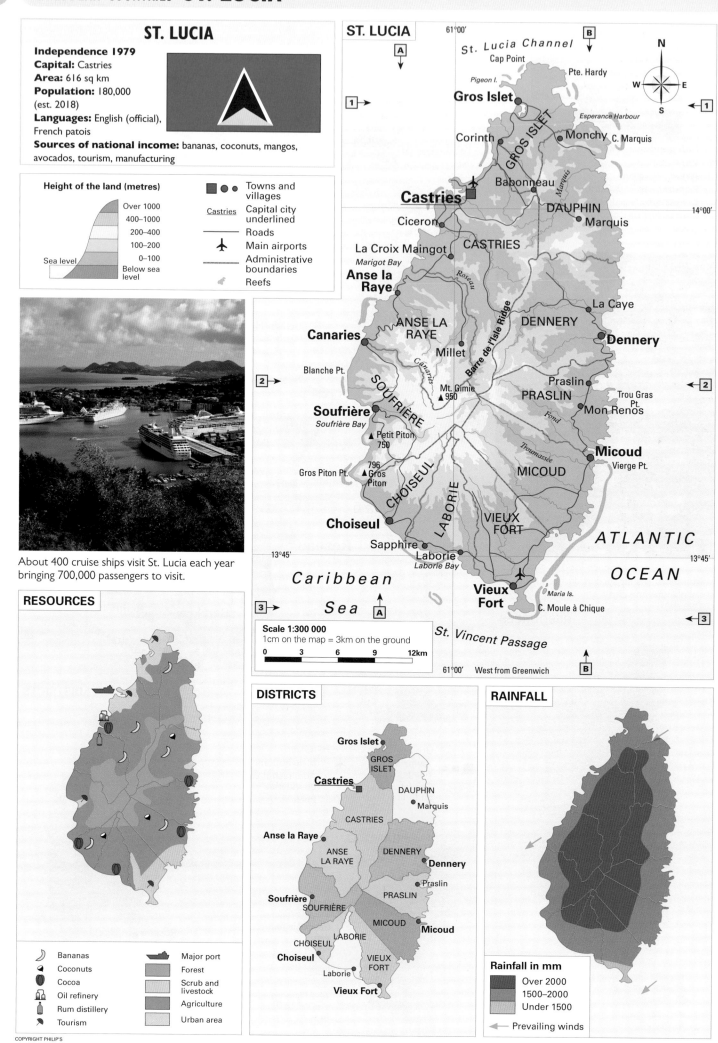

ST. LUCIA

St. Lucia Channel
Cap Point
Pte. Hardy
Pigeon I.
Gros Islet
Esperance Harbour
Corinth
Monchy
C. Marquis
Babonneau
DAUPHIN
Castries
Ciceron
Marquis
CASTRIES
La Croix Maingot
Marigot Bay
Anse la Raye
La Caye
DENNERY
ANSE LA RAYE
Millet
Dennery
Canaries
Blanche Pt.
SOUFRIÈRE
Praslin
PRASLIN
Trou Gras Pt.
Mt. Gimie ▲950
Mon Renos
Soufrière
Soufrière Bay
▲ Petit Piton 750
Micoud
Vierge Pt.
Gros Piton Pt. ▲796 Gros Piton
CHOISEUL
MICOUD
Choiseul
LABORIE
VIEUX FORT
Sapphire
Laborie
ATLANTIC
Caribbean
Laborie Bay
OCEAN
Sea
Vieux Fort
Maria Is.
C. Moule à Chique
St. Vincent Passage

Scale 1:300 000
1cm on the map = 3km on the ground
0   3   6   9   12km

61°00'   West from Greenwich

DISTRICTS

Gros Islet
GROS ISLET
Castries
DAUPHIN
Marquis
CASTRIES
Anse la Raye
ANSE LA RAYE
DENNERY
Dennery
Praslin
Soufrière
PRASLIN
SOUFRIÈRE
MICOUD
Micoud
CHOISEUL
LABORIE
Choiseul
Laborie
VIEUX FORT
Vieux Fort

RAINFALL

Rainfall in mm
Over 2000
1500–2000
Under 1500
Prevailing winds

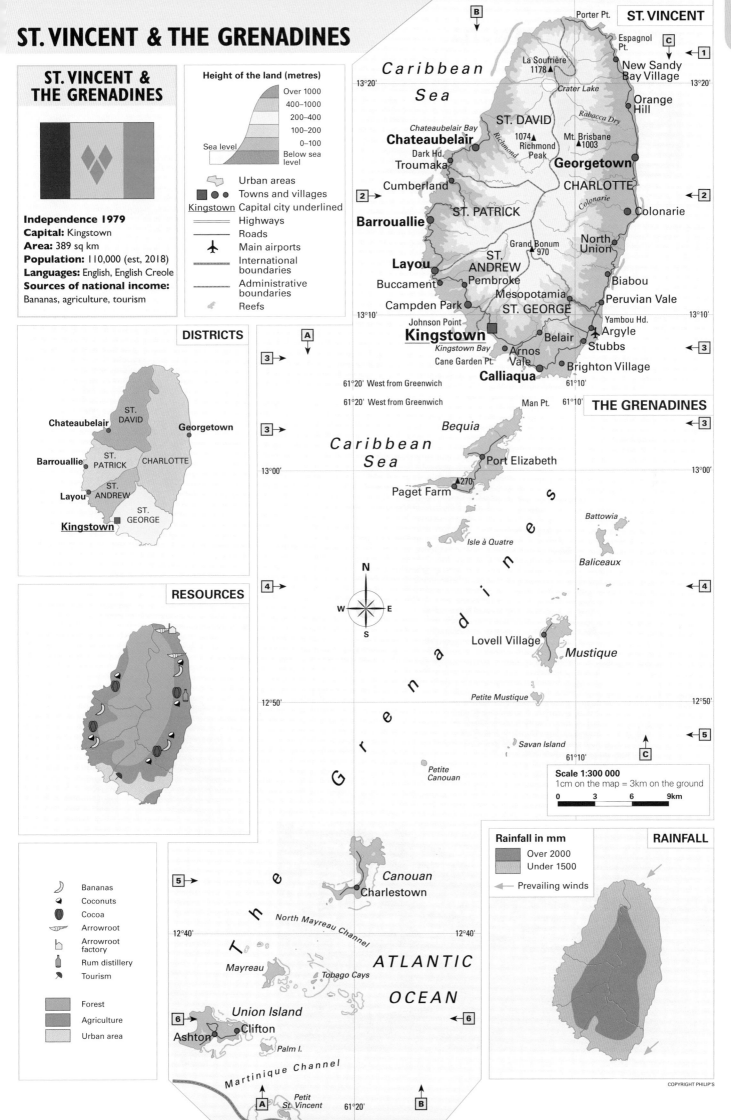

ST. VINCENT & THE GRENADINES

ST. VINCENT & THE GRENADINES

Independence 1979
Capital: Kingstown
Area: 389 sq km
Population: 110,000 (est, 2018)
Languages: English, English Creole
Sources of national income:
Bananas, agriculture, tourism

Height of the land (metres)
Over 1000
400–1000
200–400
100–200
0–100
Sea level
Below sea level

Urban areas
Towns and villages
Kingstown Capital city underlined
Highways
Roads
Main airports
International boundaries
Administrative boundaries
Reefs

ST. VINCENT

Caribbean Sea

Porter Pt.
Espagnol Pt.
New Sandy Bay Village
La Soufrière 1178▲
Crater Lake
Orange Hill
ST. DAVID
Rabacca Dry
Chateaubelair Bay
Chateaubelair
Dark Hd.
Troumaka
1074▲ Richmond Richmond Peak
Mt. Brisbane ▲1003
Georgetown
CHARLOTTE
Cumberland
Colonarie
Barrouallie
ST. PATRICK
Colonarie
North Union
Grand Bonum ▲970
Layou
ST. ANDREW
Buccament
Pembroke
Biabou
Mesopotamia
Peruvian Vale
Campden Park
ST. GEORGE
Johnson Point
Yambou Hd.
Kingstown
Belair
Argyle
Kingstown Bay
Arnos Vale
Stubbs
Cane Garden Pt.
Brighton Village
Calliaqua

61°20' West from Greenwich
61°20' West from Greenwich

DISTRICTS

Chateaubelair
ST. DAVID
Georgetown
Barrouallie
ST. PATRICK
CHARLOTTE
Layou
ST. ANDREW
Kingstown
ST. GEORGE

RESOURCES

THE GRENADINES

Caribbean Sea

Man Pt.
Bequia
Port Elizabeth
▲270
Paget Farm
Isle à Quatre
Battowia
Baliceaux
Lovell Village
Mustique
Petite Mustique
Savan Island
Petite Canouan

N
W E
S

Scale 1:300 000
1cm on the map = 3km on the ground
0   3   6   9km

RAINFALL

Rainfall in mm
Over 2000
Under 1500
Prevailing winds

Bananas
Coconuts
Cocoa
Arrowroot
Arrowroot factory
Rum distillery
Tourism

Forest
Agriculture
Urban area

Canouan
Charlestown
North Mayreau Channel
Mayreau
Tobago Cays
ATLANTIC OCEAN
Union Island
Clifton
Ashton
Palm I.
Martinique Channel
Petit St. Vincent

COPYRIGHT PHILIP'S

GRENADA

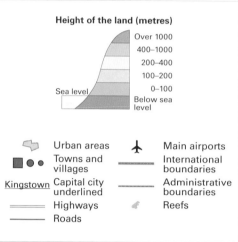

Independence: 1974
Capital: St George's
Area: 344 sq km
Population: 108,000 (est. 2018)
Languages: English (official), French patios
Sources of national income:
bananas, cocoa, nutmeg, fruit, vegetables, clothing, mace, tourism

Height of the land (metres)

Over 1000
400–1000
200–400
100–200
0–100
Below sea level
Sea level

- Urban areas
- Towns and villages
- Kingstown — Capital city underlined
- Highways
- Roads
- Main airports
- International boundaries
- Administrative boundaries
- Reefs

CARRIACOU

Same scale as Grenada

61°30'

12°30'

Gun Pt.

291 Windward

Petit St. Vincent

Petite Martinique

Hillsborough Bay

Hillsborough

Caribbean Sea

Tyrrel Bay Hermitage

Southwest Pt.

Carriacou (Grenada)

Saline I.

ATLANTIC OCEAN

Frigate I.

Large I.

61°30'

GRENADA

61°40'

Diamond I.

Les Tantes

Ronde Island

Caille I.

Caribbean Sea

N
W E
S

Tanga Langua

Sauteurs Green I.

Bedford Pt.

Grenada Bay

Victoria

ST. PATRICK

River Sallee

Tricolar

Artiste Pt.

ST. MARK

840 Tivoli

Gouyave

Mt. St. Catherine

Pearls

Grand Roy **ST. JOHN**

Concord

Great River Bay

Great River

ST. ANDREW

Grenville

Telescope Pt.

Grenville Bay

Birch Grove

Moliniere Pt.

702 *St. Francis*

Mt. Sinai

ATLANTIC

Great Bacolet Bay

OCEAN

St. George's

ST. GEORGE

St. Pauls

ST. DAVID

St. Davids

Grand Anse Bay

Morne Rouge

Corinth

Requin Bay

Pt. Salines

Lance aux Epines

Hog I.

Pt. of Fort Jeudy

Calivigny I.

Glover I. Prickly Pt.

West from Greenwich 61°40'

12°10'

12°00'

Scale 1:300 000
1cm on the map = 3km on the ground
0 3 6 9 12km

RESOURCES

- 🐟 Fishing
- Bananas
- Coconuts
- Cocoa
- Nutmeg
- Sugar
- Chocolate factory
- Nutmeg factory
- Rum distillery
- Tourism
- Forest
- Agriculture
- Urban area

RAINFALL

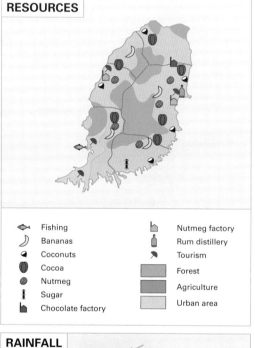

Rainfall in mm
- Over 3000
- 2000–3000
- 1500–2000
- 1000–1500
- Under 1000
- ← Prevailing winds

DISTRICTS

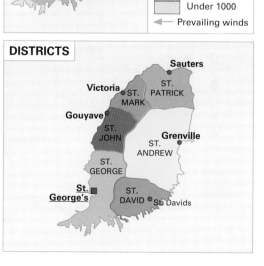

Sauters

Victoria ST. PATRICK

ST. MARK

Gouyave

ST. JOHN

Grenville

ST. ANDREW

ST. GEORGE

St. George's

ST. DAVID St. Davids

The picturesque Carenage of Grenada's capital St Georges. Increasingly Grenada has become a centre for yachting and the harbour hosts a marina.

There are six islands of the Caribbean Netherlands. Three are countries of the Kingdom of the Netherlands, namely Curaçao, Aruba and Sint Maarten. The other three are administrative parts of the country of the Netherlands, namely Bonaire, Sint Eustatius and Saba.

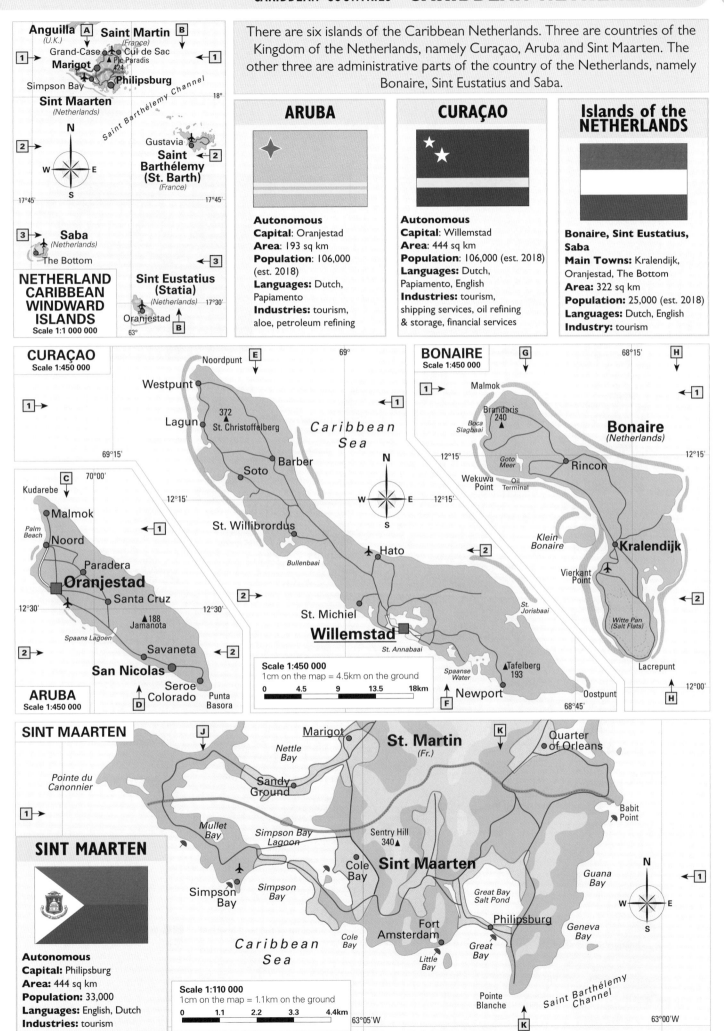

ARUBA

Autonomous
Capital: Oranjestad
Area: 193 sq km
Population: 106,000 (est. 2018)
Languages: Dutch, Papiamento
Industries: tourism, aloe, petroleum refining

CURAÇAO

Autonomous
Capital: Willemstad
Area: 444 sq km
Population: 106,000 (est. 2018)
Languages: Dutch, Papiamento, English
Industries: tourism, shipping services, oil refining & storage, financial services

Islands of the NETHERLANDS

Bonaire, Sint Eustatius, Saba
Main Towns: Kralendijk, Oranjestad, The Bottom
Area: 322 sq km
Population: 25,000 (est. 2018)
Languages: Dutch, English
Industry: tourism

NETHERLAND CARIBBEAN WINDWARD ISLANDS
Scale 1:1 000 000

CURAÇAO
Scale 1:450 000

ARUBA
Scale 1:450 000

BONAIRE
Scale 1:450 000

Scale 1:450 000
1cm on the map = 4.5km on the ground

0   4.5   9   13.5   18km

SINT MAARTEN

SINT MAARTEN

Autonomous
Capital: Philipsburg
Area: 444 sq km
Population: 33,000
Languages: English, Dutch
Industries: tourism

Scale 1:110 000
1cm on the map = 1.1km on the ground

0   1.1   2.2   3.3   4.4km

COPYRIGHT PHILIP'S

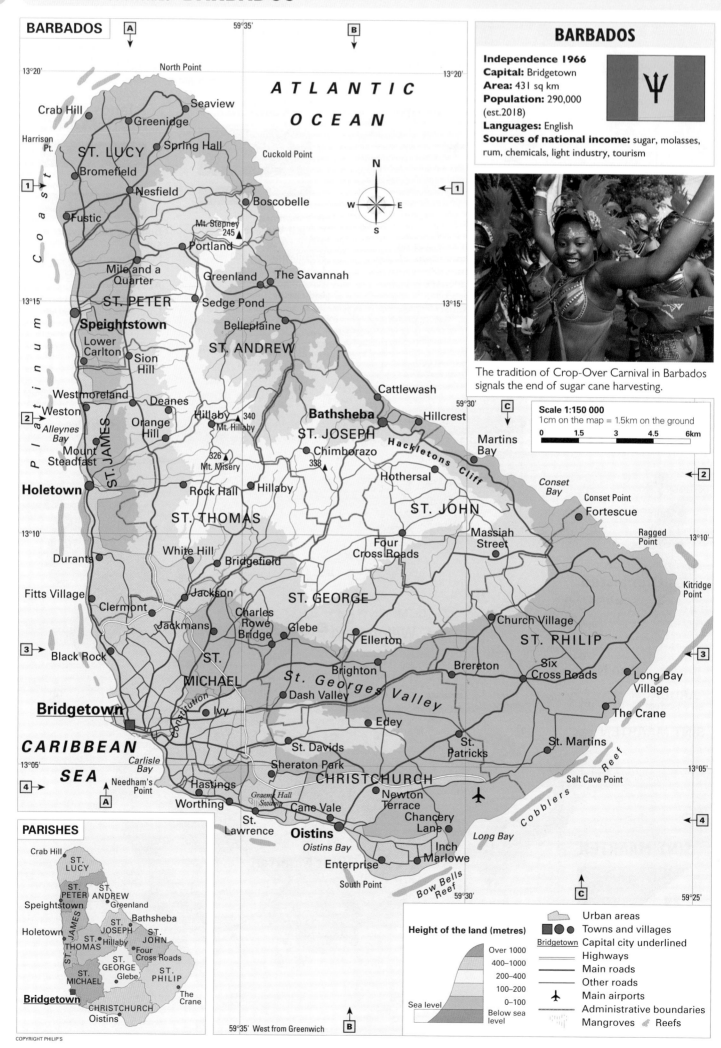

BARBADOS

BARBADOS

Independence 1966
Capital: Bridgetown
Area: 431 sq km
Population: 290,000
(est.2018)
Languages: English
Sources of national income: sugar, molasses, rum, chemicals, light industry, tourism

The tradition of Crop-Over Carnival in Barbados signals the end of sugar cane harvesting.

Scale 1:150 000
1cm on the map = 1.5km on the ground
0   1.5   3   4.5   6km

Map labels

ATLANTIC OCEAN

North Point
Crab Hill
Seaview
Greenidge
Harrison Pt.
ST. LUCY
Spring Hall
Bromefield
Nesfield
Cuckold Point
Fustic
Boscobelle
Mt. Stepney 245
Portland
Mile and a Quarter
Greenland
The Savannah
ST. PETER
Sedge Pond
Speightstown
Belleplaine
Lower Carlton
Sion Hill
ST. ANDREW
Westmoreland
Deanes
Cattlewash
Weston
Hillaby 340
Bathsheba
Hillcrest
Alleynes Bay
Orange Hill
Mt. Hillaby
ST. JOSEPH
Martins Bay
Mount Steadfast
326 Mt. Misery
338
Chimborazo
Hackletons Cliff
Holetown
Rock Hall
Hillaby
Hothersal
Conset Bay
ST. THOMAS
ST. JOHN
Conset Point
Fortescue
White Hill
Bridgefield
Four Cross Roads
Massiah Street
Ragged Point
Durants
Jackson
ST. GEORGE
Kitridge Point
Fitts Village
Clermont
Charles Rowe Bridge
Glebe
Church Village
Jackmans
Ellerton
ST. PHILIP
Black Rock
ST. MICHAEL
Brighton
Brereton
Six Cross Roads
Long Bay Village
St. Georges Valley
Dash Valley
Bridgetown
Ivy
Edey
The Crane
CARIBBEAN
St. Davids
St. Patricks
St. Martins
SEA
Carlisle Bay
Sheraton Park
CHRISTCHURCH
Salt Cave Point
Needham's Point
Hastings
Newton Terrace
Worthing
Graeme Hall Swamp
Cane Vale
Chancery Lane
St. Lawrence
St. Martins
Long Bay
Oistins
Inch Marlowe
Oistins Bay
Enterprise
South Point
Bow Bells Reef
Cobblers Reef

PARISHES

Crab Hill
ST. LUCY
ST. PETER
ST. ANDREW
Speightstown
Greenland
ST. JAMES
Bathsheba
Holetown
ST. JOSEPH
ST. JOHN
ST. THOMAS
Hillaby
Four Cross Roads
ST. GEORGE
ST. PHILIP
Glebe
ST. MICHAEL
The Crane
Bridgetown
CHRISTCHURCH
Oistins

59°35' West from Greenwich

Legend

Height of the land (metres)
Over 1000
400–1000
200–400
100–200
0–100
Sea level
Below sea level

Urban areas
Towns and villages
Bridgetown Capital city underlined
Highways
Main roads
Other roads
Main airports
Administrative boundaries
Mangroves   Reefs

RAINFALL

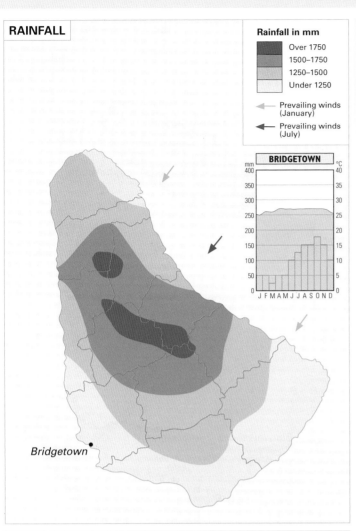

Rainfall in mm	
	Over 1750
	1500–1750
	1250–1500
	Under 1250

← Prevailing winds (January)

← Prevailing winds (July)

BRIDGETOWN

Bridgetown

RESOURCES

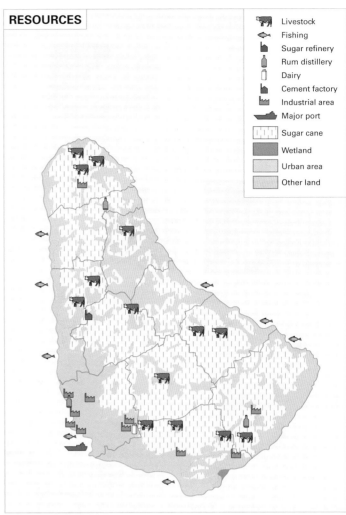

	Livestock
	Fishing
	Sugar refinery
	Rum distillery
	Dairy
	Cement factory
	Industrial area
	Major port
	Sugar cane
	Wetland
	Urban area
	Other land

POPULATION

URBAN POPULATION

■ 50,000–100,000 people

● Other towns

AGE DISTRIBUTION PYRAMID (2015)

Years Old

MALES   FEMALES

80+
70–79
60–69
50–59
40–49
30–39
20–29
10–19
0–9

15 10 5 %   % 5 10 15

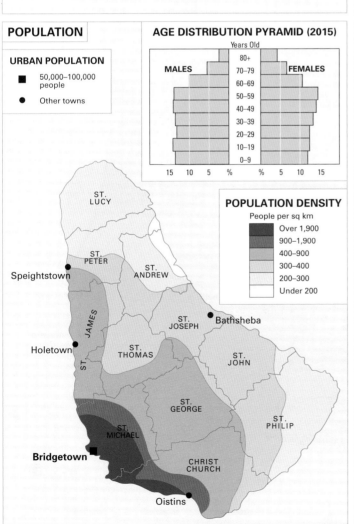

ST. LUCY
ST. PETER
ST. ANDREW
Speightstown
ST. JAMES
ST. JOSEPH
Bathsheba
Holetown
ST. THOMAS
ST. JOHN
ST. ANDREW
ST. GEORGE
ST. MICHAEL
Bridgetown
CHRIST CHURCH
ST. PHILIP
Oistins

POPULATION DENSITY

People per sq km

	Over 1,900
	900–1,900
	400–900
	300–400
	200–300
	Under 200

TOURISM

	Caves		Watersports
	Beach/bathing areas		Botanic garden
	Plantation house		Golf course
	Bird sanctuary/ wildlife reserve		Museum
	Deep-sea fishing		National Park
			Main tourist areas

TOURIST ATTRACTIONS

1. Blackmans Bridge
2. Bridgetown & Garrison
3. Cherry Tree Hill
4. Christ Church
5. Codrington College
6. Cotton Tower Signal Station
7. Folkestone Underwater Park & Marine Museum
8. Foursquare Rum Factory & Heritage Park
9. Government House
10. Gun Hill Tower
11. Kensington Oval
12. Malibu Visitor Centre
13. Morgan Lewis Mill
14. Mount Gay Visitor Centre
15. National Stadium
16. St John's Church
17. Sam Lord's Castle
18. Turners Hall Woods

Speightstown

Bathsheba

Holetown

Bridgetown

Hastings

Grantley Adams International Airport

Oistins

COPYRIGHT PHILIP'S

TRINIDAD & TOBAGO

Independence: 1962
Capital: Port of Spain
Area: 5,128 sq km
Population: 1,380,000 (est. 2018)
Languages: English (official), Creole English, Caribbean Hindi, Spanish, Chinese
Sources of national income: beverages, cocoa, petroleum, chemicals, natural gas, steel products, tourism
National birds: scarlet ibis and cocrico

National Flower: Chaconia   Coat of Arms

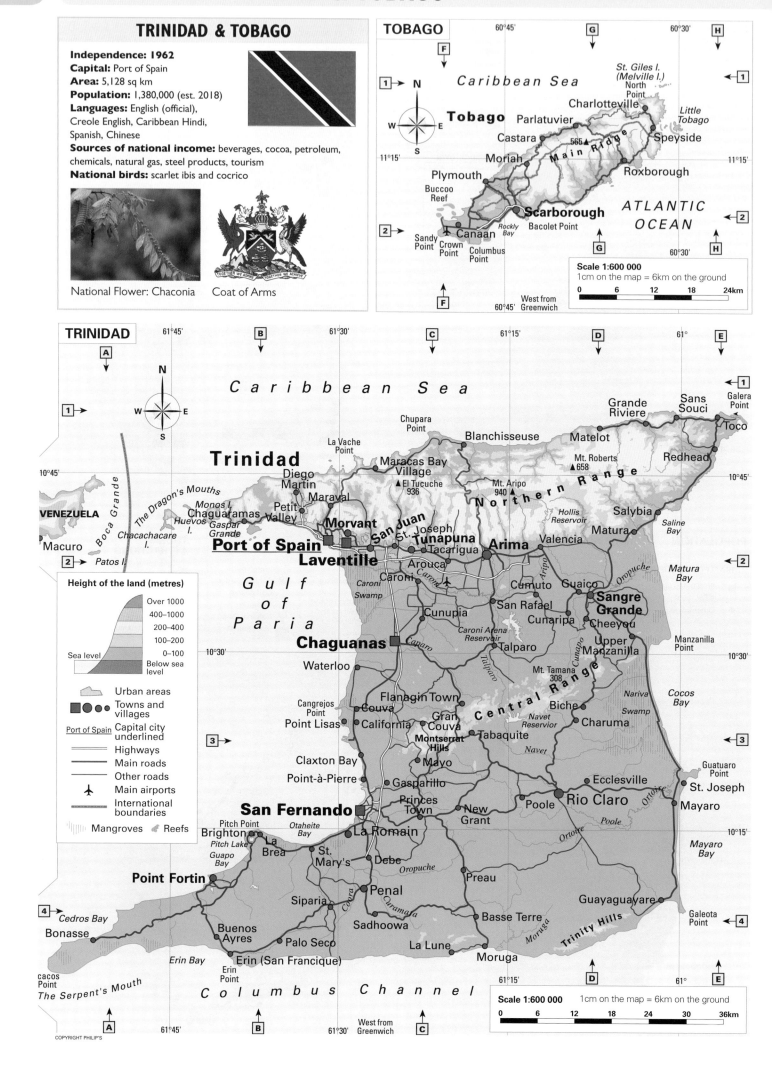

Height of the land (metres)
Over 1000
400–1000
200–400
100–200
0–100
Sea level
Below sea level

Urban areas
Towns and villages
Port of Spain — Capital city underlined
Highways
Main roads
Other roads
Main airports
International boundaries
Mangroves   Reefs

Scale 1:600 000
1cm on the map = 6km on the ground
0 6 12 18 24 30 36km

REGIONS AND BOROUGHS

- ■ City
- ● Town
- SIPARIA Region
- ARIMA Borough
- **PORT OF SPAIN** City

CITIES, TOWNS & BOROUGHS

TOBAGO (Ward)

Scarborough

TOBAGO

ST. DAVID

ST. GEORGE

ST. ANDREW

CARONI

NARIVA

VICTORIA

MAYARO

ST. PATRICK

HISTORIC COUNTIES

DIEGO MARTIN

Petit Valley

Port of Spain

PORT OF SPAIN

Laventille

SAN JUAN/ LAVENTILLE

Tunapuna

ARIMA

Arima

TUNAPUNA/ PIARCO

SANGRE GRANDE

Sangre Grande

CHAGUANAS

Chaguanas

Couva

COUVA/ TABAQUITE/ TALPARO

Rio Claro

SAN FERNANDO

San Fernando

Princes Town

MAYARO/ RIO CLARO

Point Fortin

POINT FORTIN

PENAL/ DEBE

Penal

Siparia

PRINCES TOWN

SIPARIA

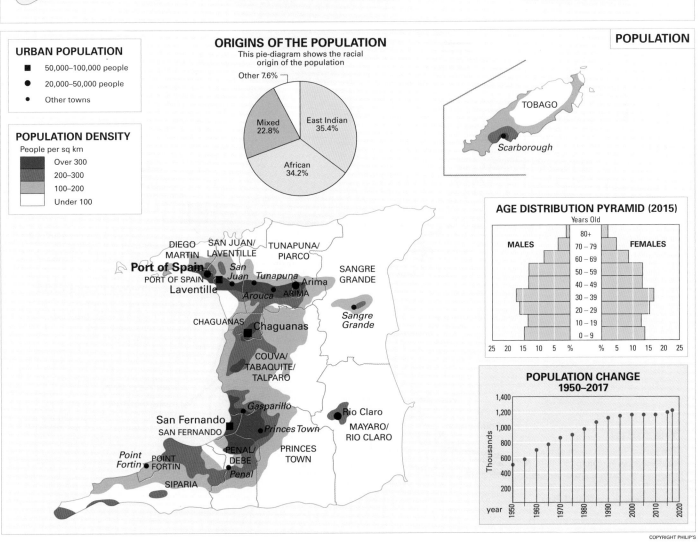

POPULATION

URBAN POPULATION

- ■ 50,000–100,000 people
- ● 20,000–50,000 people
- ● Other towns

POPULATION DENSITY

People per sq km
- Over 300
- 200–300
- 100–200
- Under 100

ORIGINS OF THE POPULATION

This pie-diagram shows the racial origin of the population

- Other 7.6%
- East Indian 35.4%
- African 34.2%
- Mixed 22.8%

TOBAGO

Scarborough

DIEGO MARTIN

SAN JUAN/ LAVENTILLE

TUNAPUNA/ PIARCO

Port of Spain

PORT OF SPAIN

Laventille

San Juan

Arouca

Tunapuna

ARIMA

Arima

SANGRE GRANDE

Sangre Grande

CHAGUANAS

Chaguanas

COUVA/ TABAQUITE/ TALPARO

Gasparillo

San Fernando

SAN FERNANDO

Princes Town

Rio Claro

MAYARO/ RIO CLARO

Point Fortin

POINT FORTIN

PENAL/ DEBE

Penal

PRINCES TOWN

SIPARIA

AGE DISTRIBUTION PYRAMID (2015)

Years Old

MALES — FEMALES

| 80+ |
| 70 – 79 |
| 60 – 69 |
| 50 – 59 |
| 40 – 49 |
| 30 – 39 |
| 20 – 29 |
| 10 – 19 |
| 0 – 9 |

25 20 15 10 5 % % 5 10 15 20 25

POPULATION CHANGE 1950–2017

Thousands (1,400 / 1,200 / 1,000 / 800 / 600 / 400 / 200)

year 1950 1960 1970 1980 1990 2000 2010 2020

CLIMATE

PIARCO

	J	F	M	A	M	J	J	A	S	O	N	D
Daily hours of sunshine	7	8	8	8	8	6	7	7	7	7	7	7
Humidity (% at 5pm)	70	67	64	65	70	75	75	76	76	78	79	76
Days with thunderstorms	0	0	0	0	1	3	5	6	7	6	3	1
Days with rain	19	16	15	14	20	24	23	22	19	20	21	21
Months of the year	J	F	M	A	M	J	J	A	S	O	N	D

CROWN POINT

	J	F	M	A	M	J	J	A	S	O	N	D
Daily hours of sunshine	7	8	8	8	8	6	7	7	7	7	7	7
Humidity (% at 5pm)	74	71	70	72	74	78	78	78	77	80	82	78
Days with thunderstorms	0	0	0	0	1	1	2	2	3	2	1	0
Days with rain	11	9	8	7	10	14	14	14	11	14	17	12
Months of the year	J	F	M	A	M	J	J	A	S	O	N	D

CLIMATE GRAPHS

Name of place — **PORT OF SPAIN**
Average annual rainfall — Rainfall 1770mm
Average monthly temperature
Average monthly rainfall
Months of the year

PIARCO — Rainfall 1723mm

CROWN POINT — Rainfall 1498mm

ANNUAL RAINFALL AND PREVAILING WIND

Rainfall in mm
- Over 2500
- 2000–2500
- 1750–2000
- 1500–1750
- Under 1500
- → Prevailing winds (January)
- → Prevailing winds (July)

Port of Spain
Piarco
Crown Point

PLACES

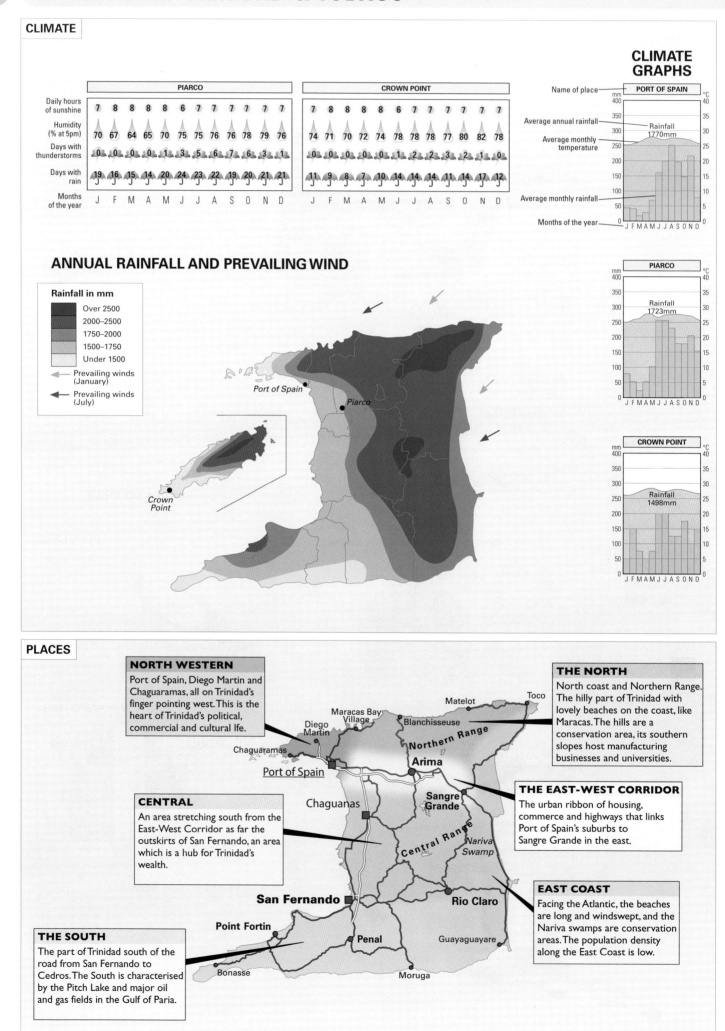

NORTH WESTERN
Port of Spain, Diego Martin and Chaguaramas, all on Trinidad's finger pointing west. This is the heart of Trinidad's political, commercial and cultural life.

THE NORTH
North coast and Northern Range. The hilly part of Trinidad with lovely beaches on the coast, like Maracas. The hills are a conservation area, its southern slopes host manufacturing businesses and universities.

CENTRAL
An area stretching south from the East-West Corridor as far the outskirts of San Fernando, an area which is a hub for Trinidad's wealth.

THE EAST-WEST CORRIDOR
The urban ribbon of housing, commerce and highways that links Port of Spain's suburbs to Sangre Grande in the east.

EAST COAST
Facing the Atlantic, the beaches are long and windswept, and the Nariva swamps are conservation areas. The population density along the East Coast is low.

THE SOUTH
The part of Trinidad south of the road from San Fernando to Cedros. The South is characterised by the Pitch Lake and major oil and gas fields in the Gulf of Paria.

Diego Martin, Maracas Bay Village, Chaguaramas, Port of Spain, Chaguanas, Matelot, Blanchisseuse, Toco, Northern Range, Arima, Sangre Grande, Central Range, Nariva Swamp, San Fernando, Rio Claro, Point Fortin, Penal, Guayaguayare, Bonasse, Moruga

AGRICULTURE

AGRICULTURAL PRODUCE
Selected produce in thousand tonnes (up to 2003)

Chicken meat

Coconuts

Citrus fruit

Dasheen

Cows milk

Bananas

Casava

Rice

Pineapple

Cocoa, beans

2014
2004

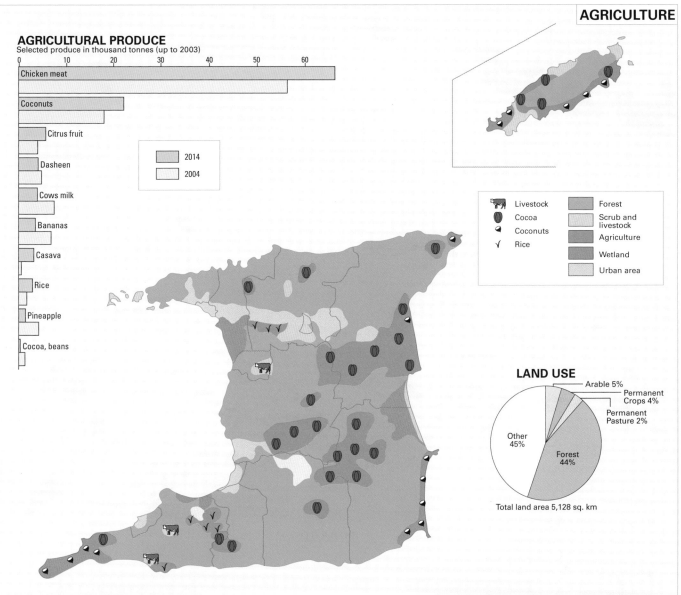

Livestock
Cocoa
Coconuts
Rice

Forest
Scrub and livestock
Agriculture
Wetland
Urban area

LAND USE

Arable 5%
Permanent Crops 4%
Permanent Pasture 2%
Other 45%
Forest 44%

Total land area 5,128 sq. km

ROLE OF AGRICULTURE IN THE ECONOMY
1966 - 2006

$ Trinidad & Tobago (million)

14,000

12,000

10,000

8,000

6,000

4,000

2,000

Domestic agriculture
Sugar industry
Manufacturing
Finance, insurance & real estate

1966 1970 1974 1978 1982 1986 1990 1994 1998 2002 2006
Year

Agriculture has not kept pace with the growth of other sectors of Trinidad and Tobago's economy.

Trinidad's output of milk and beef has been increasing, but the country still needs to import as well.

Fresh coconuts for sale, Port of Spain, Trinidad

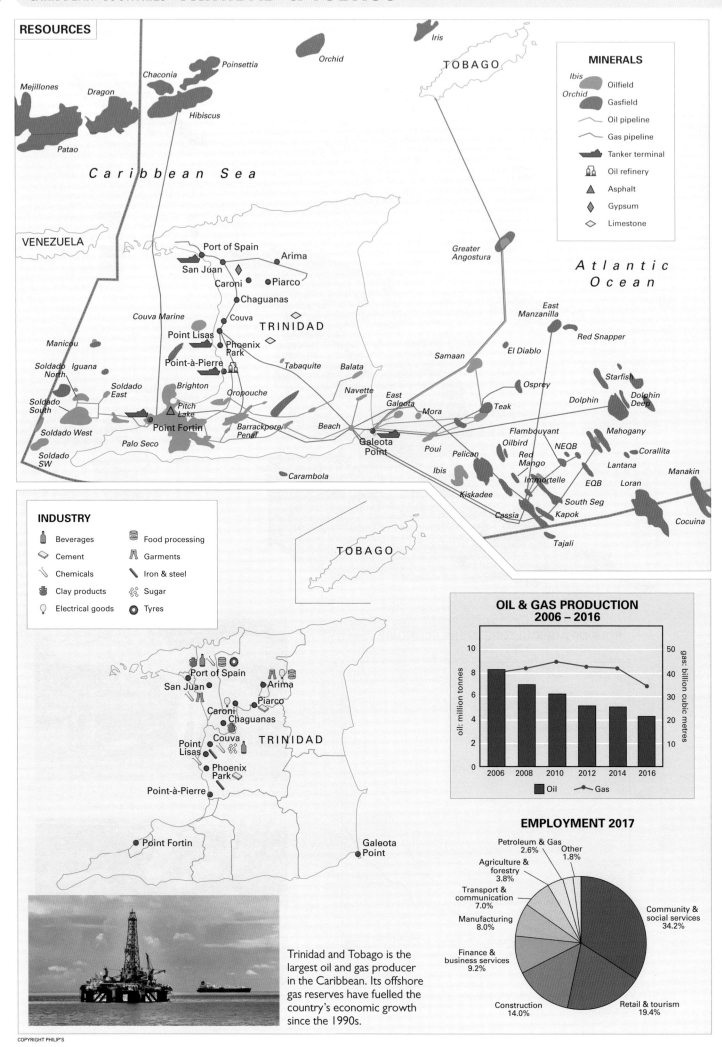

RESOURCES

Iris

Orchid

TOBAGO

Mejillones

Dragon

Chaconia

Poinsettia

Hibiscus

Patao

C a r i b b e a n   S e a

VENEZUELA

MINERALS

Ibis

Orchid

- Oilfield
- Gasfield
- Oil pipeline
- Gas pipeline
- Tanker terminal
- Oil refinery
- Asphalt
- Gypsum
- Limestone

Port of Spain
Arima
San Juan
Caroni
Piarco
Chaguanas
Couva
TRINIDAD
Couva Marine
Point Lisas
Phoenix Park
Point-à-Pierre
Tabaquite
Balata
Manicou
Brighton
Oropouche
Navette
Soldado North
Iguana
Soldado East
Pitch Lake
Point Fortin
Barrackpore/Penal
Beach
Galeota Point
Soldado South
Soldado West
Palo Seco
Soldado SW
Carambola

Greater Angostura

A t l a n t i c   O c e a n

East Manzanilla
Red Snapper
El Diablo
Samaan
Osprey
Starfish
Dolphin
Dolphin Deep
East Galeota
Mora
Teak
Mahogany
Flambouyant
Oilbird
NEQB
Corallita
Poui
Red Mango
Lantana
Pelican
Immortelle
EQB
Loran
Manakin
Ibis
Kiskadee
South Seg
Kapok
Cocuina
Cassia
Tajali

INDUSTRY

- Beverages
- Cement
- Chemicals
- Clay products
- Electrical goods
- Food processing
- Garments
- Iron & steel
- Sugar
- Tyres

TOBAGO

Port of Spain
San Juan
Arima
Piarco
Caroni
Chaguanas
Couva
TRINIDAD
Point Lisas
Phoenix Park
Point-à-Pierre

Point Fortin
Galeota Point

OIL & GAS PRODUCTION 2006 – 2016

oil: million tonnes

gas: billion cubic metres

| | 2006 | 2008 | 2010 | 2012 | 2014 | 2016 |

■ Oil ◆ Gas

EMPLOYMENT 2017

- Petroleum & Gas 2.6%
- Other 1.8%
- Agriculture & forestry 3.8%
- Transport & communication 7.0%
- Manufacturing 8.0%
- Finance & business services 9.2%
- Construction 14.0%
- Retail & tourism 19.4%
- Community & social services 34.2%

Trinidad and Tobago is the largest oil and gas producer in the Caribbean. Its offshore gas reserves have fuelled the country's economic growth since the 1990s.

TOURISM

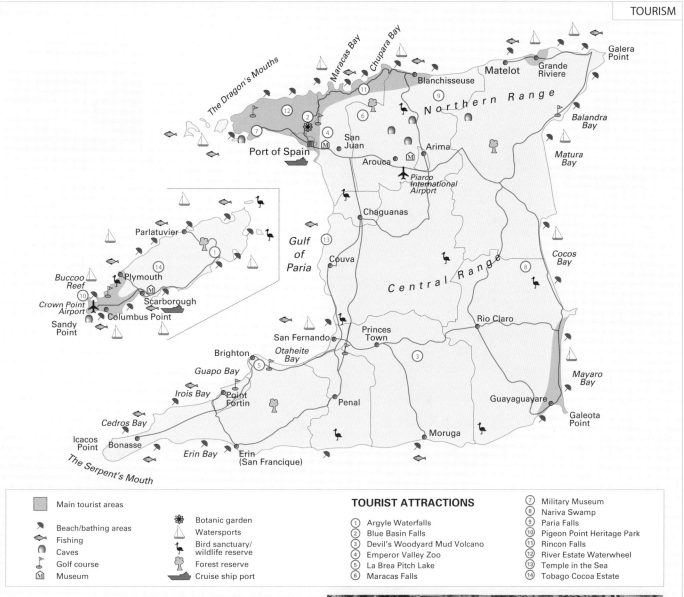

The Dragon's Mouths
Maracas Bay
Chupara Bay
Galera Point
Matelot
Grande Riviere
Blanchisseuse
Northern Range
Balandra Bay
Port of Spain
San Juan
Arima
Matura Bay
Arouca
Piarco International Airport
Chaguanas
Gulf of Paria
Couva
Central Range
Cocos Bay
Parlatuvier
Buccoo Reef
Plymouth
Crown Point Airport
Scarborough
Columbus Point
Sandy Point
Rio Claro
Princes Town
San Fernando
Otaheite Bay
Brighton
Guapo Bay
Mayaro Bay
Irois Bay
Point Fortin
Penal
Guayaguayare
Cedros Bay
Galeota Point
Icacos Point
Bonasse
Erin Bay
Erin (San Francique)
Moruga
The Serpent's Mouth

▨	Main tourist areas	❀	Botanic garden
⛱	Beach/bathing areas	⛵	Watersports
🐟	Fishing	🦩	Bird sanctuary/ wildlife reserve
⋒	Caves	🌳	Forest reserve
⛳	Golf course	🚢	Cruise ship port
Ⓜ	Museum		

TOURIST ATTRACTIONS

① Argyle Waterfalls
② Blue Basin Falls
③ Devil's Woodyard Mud Volcano
④ Emperor Valley Zoo
⑤ La Brea Pitch Lake
⑥ Maracas Falls

⑦ Military Museum
⑧ Nariva Swamp
⑨ Paria Falls
⑩ Pigeon Point Heritage Park
⑪ Rincon Falls
⑫ River Estate Waterwheel
⑬ Temple in the Sea
⑭ Tobago Cocoa Estate

STOPOVER AND CRUISE PASSENGER ARRIVALS

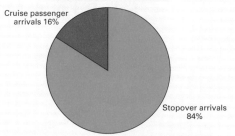

Cruise passenger arrivals 16%

Stopover arrivals 84%

Total visitor arrivals in 2016 488,000

ORIGIN OF STOPOVER TOURISTS

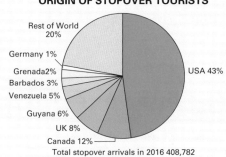

Rest of World 20%
Germany 1%
Grenada 2%
Barbados 3%
Venezuela 5%
Guyana 6%
UK 8%
Canada 12%
USA 43%

Total stopover arrivals in 2016 408,782

Trinidad & Tobago created steel band music. It's a big attraction for visitors throughout the year, and orchestras are invited to perform in many countries.

Nariva, in eastern Trinidad, is the largest fresh-water wetland in the Caribbean, a place to meet up with monkeys, toucans, macaws, iguanas and other wildlife in a natural protected habitat.

LOCATOR MAP

PUERTO RICO
ANTIGUA AND BARBUDA
GUADELOUPE
DOMINICA
MARTINIQUE ST. LUCIA
BARBADOS
GRENADA
TRINIDAD AND TOBAGO
VENEZUELA
GUYANA
SURINAME
FRENCH GUIANA
BRAZIL

GUYANA

Independence 1966
Capital: Georgetown
Area: 214,970 sq km
Population: 747,000 (census 2012)
Languages: English (official), Guyanese Creole, Amerindian dialects, Creole, Indian languages, Chinese
Sources of national income: sugar, rice, shrimps, gold, bauxite, timber

VENEZUELA A

ATLANTIC OCEAN

N W E S

Mabaruma
Mabaruma
Port Kaituma
Baramanni
BARIMA-WAINI
Bochinche
Matthews Ridge
Kokerite
Charity
POMEROON-SUPERNAAM
Anna Regina
Vreed-en-Hoop
Parika
Georgetown
ESSEQUIBO ISLANDS WEST DEMERARA
Helena
Mahaicony
Fort Wellington
New Amsterdam
Rose Hall
Rosignol
MAHAICA-BERBICE
Everton
Nieuw-Nickerie
Totness
Arimu Mine
Peter's Mine
Bartica
DEMERARA-MAHAICA
CUYUNI-MAZARUNI
Rockstone
Linden
Corriverton
Wageningen
Boskamp
Kamarang
Issano
UPPER
Ituni
EAST BERBICE-CORENTYNE
Orealla
Apoera
Merume Mts.
G U Y A N A
Potaro Landing
Kaieteur Falls
Tumatumari
DEMERARA-
Kwakwani
Epira
Pakaraima
Mt. Roraima 2810
Mahdia
POTARO-SIPARUNI
BERBICE
Davis Dam
Mountains
Sierra Pacaraima
B R A Z I L
Kurupukari
Corentyne
Kabalebo Dam
S U R I N A M E
Wilhelmina Geb.
Julianatop 1230
Annai
Apoteri
Yupukarri
Pirara
Bonfim
Letham
Kanuku Mountains
Iliwa
Essequibo
Käyser Geb.
EAST BERBICE-CORENTYNE
UPPER TAKUTU-
Dadanawa
New
Oronoque Camp
Rupununi
UPPER ESSEQUIBO
Aishalton
Achiwib
Marudi Mts.
Amuku Mts.
Biloku
Kamoa Mts.
734
B R A Z I L

60° West from Greenwich 58°

Height of the land (metres)
Over 4000
2000–4000
1000–2000
400–1000
200–400
0–200
Sea level
Below sea level

Urban areas
Towns and villages
Georgetown Capital city underlined
Roads
Main airports
International boundaries
Administrative boundaries
Mangroves

Scale 1:3 500 000 1cm on the map = 35km on the ground
0   35   70   105   140   175   210km

COPYRIGHT PHILIP'S

CLIMATE

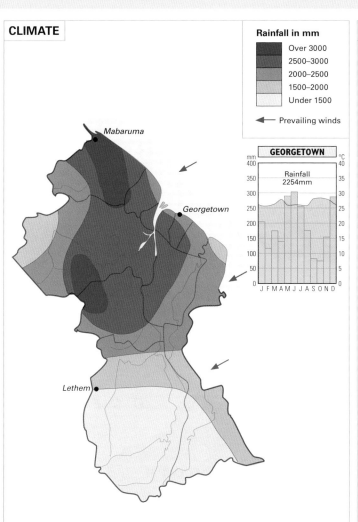

Rainfall in mm

Over 3000
2500–3000
2000–2500
1500–2000
Under 1500

← Prevailing winds

GEORGETOWN

Rainfall
2254mm

Mabaruma
Georgetown
Lethem

RESOURCES

◆ Diamonds
◇ Gold
▢ Manganese
▟ Sugar refinery
▟▟ Bauxite/
alumina plant
Glass and
clay tile
Glass sand
Bauxite areas
← Bauxite exports
Oilfield

POPULATION

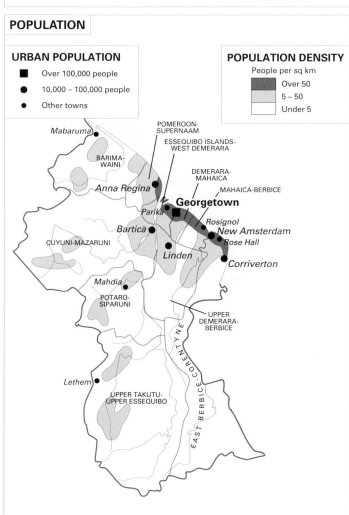

URBAN POPULATION

■ Over 100,000 people
● 10,000 – 100,000 people
● Other towns

POPULATION DENSITY

People per sq km

Over 50
5 – 50
Under 5

Mabaruma
POMEROON-
SUPERNAAM
ESSEQUIBO ISLANDS-
WEST DEMERARA
BARIMA-
WAINI
DEMERARA-
MAHAICA
Anna Regina
MAHAICA-BERBICE
Parika
Georgetown
Bartica
Rosignol
New Amsterdam
Rose Hall
CUYUNI-MAZARUNI
Linden
Corriverton
Mahdia
POTARO-
SIPARUNI
UPPER
DEMERARA-
BERBICE
EAST BERBICE-CORENTYNE
Lethem
UPPER TAKUTU-
UPPER ESSEQUIBO

AGRICULTURE & FORESTRY

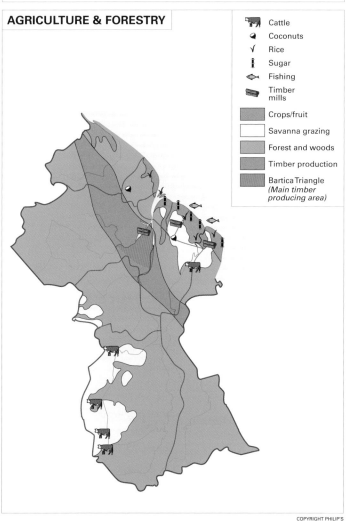

🐂 Cattle
🥥 Coconuts
Ⴅ Rice
Ⅰ Sugar
🐟 Fishing
Timber
mills

Crops/fruit
Savanna grazing
Forest and woods
Timber production
Bartica Triangle
(Main timber
producing area)

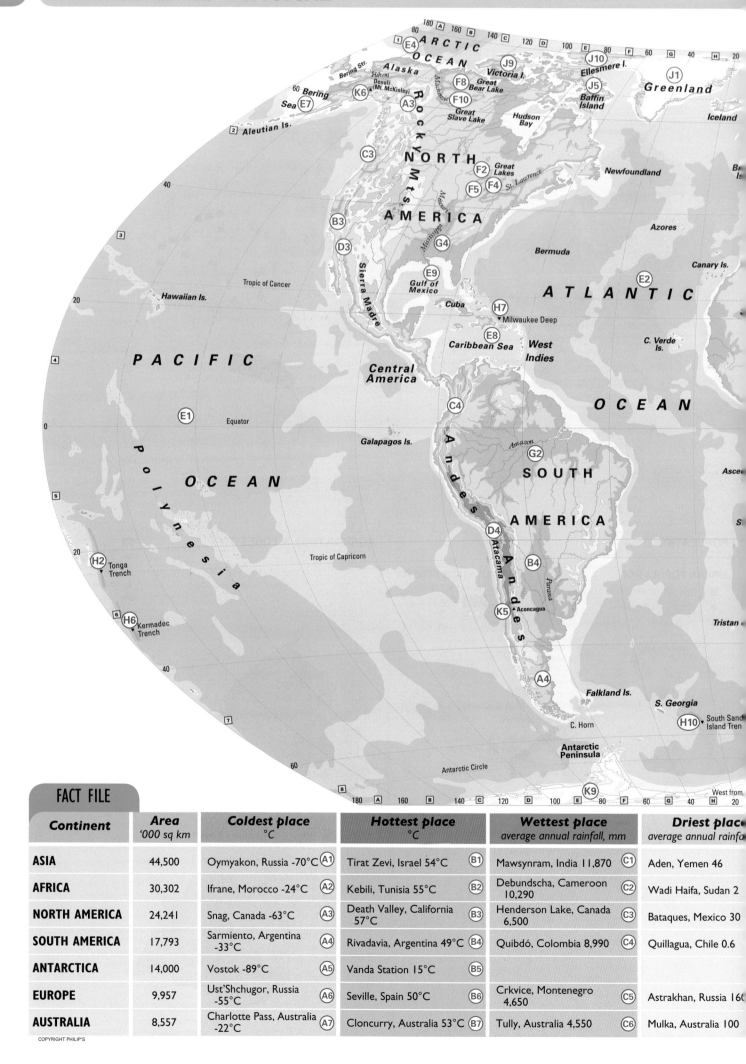

FACT FILE					
Continent	**Area** '000 sq km	**Coldest place** °C	**Hottest place** °C	**Wettest place** average annual rainfall, mm	**Driest place** average annual rainfall
ASIA	44,500	Oymyakon, Russia -70°C (A1)	Tirat Zevi, Israel 54°C (B1)	Mawsynram, India 11,870 (C1)	Aden, Yemen 46
AFRICA	30,302	Ifrane, Morocco -24°C (A2)	Kebili, Tunisia 55°C (B2)	Debundscha, Cameroon 10,290 (C2)	Wadi Haifa, Sudan 2
NORTH AMERICA	24,241	Snag, Canada -63°C (A3)	Death Valley, California 57°C (B3)	Henderson Lake, Canada 6,500 (C3)	Bataques, Mexico 30
SOUTH AMERICA	17,793	Sarmiento, Argentina -33°C (A4)	Rivadavia, Argentina 49°C (B4)	Quibdó, Colombia 8,990 (C4)	Quillagua, Chile 0.6
ANTARCTICA	14,000	Vostok -89°C (A5)	Vanda Station 15°C (B5)		
EUROPE	9,957	Ust'Shchugor, Russia -55°C (A6)	Seville, Spain 50°C (B6)	Crkvice, Montenegro 4,650 (C5)	Astrakhan, Russia 160
AUSTRALIA	8,557	Charlotte Pass, Australia -22°C (A7)	Cloncurry, Australia 53°C (B7)	Tully, Australia 4,550 (C6)	Mulka, Australia 100

Equatorial Scale 1:100 000 000
This distance is 4,000 kilometres

Height of the land (metres)
Over 6000
4000–6000
2000–4000
1000–2000
200–1000
0–200
Sea level
Below sea level

Depth of the sea (metres)
0–200
200–4000
4000–8000
Over 8000

COPYRIGHT PHILIP'S

Largest seas '000 sq km		Largest lakes '000 sq km		Longest rivers kilometres		Deepest trenches metres		Largest islands '000 sq km		Highest peaks metres	
Ocean 155,557	E1	Caspian Sea 371	F1	Nile 6,695	G1	Mariana Trench 11,022	H1	Greenland 2,176	J1	Himalayas: Mt. Everest 8,850	K1
Ocean 76,762	E2	Lake Superior 82	F2	Amazon 6,450	G2	Tonga Trench 10,822	H2	New Guinea 821	J2	Karakoram Range: K2 8,611	K2
Ocean 68,556	E3	Lake Victoria 68	F3	Yangtse 6,380	G3	Japan Trench 10,554	H3	Borneo 744	J3	Pamirs: Ismoil Somoni Pk. 7,495	K3
Ocean 14,056	E4	Lake Huron 60	F4	Mississippi-Missouri 5,971	G4	Kuril Trench 10,542	H4	Madagascar 587	J4	Tian Shan: Pik Pobedy 7,439	K4
ranean Sea 2,966	E5	Lake Michigan 58	F5	Yenisey-Angara 5,550	G5	Mindanao Trench 10,497	H5	Baffin Island 508	J5	Andes: Aconcagua 6,962	K5
hina Sea 2,318	E6	Lake Tanganyika 33	F6	Hwang Ho 5,464	G6	Kermadec Trench 10,047	H6	Sumatra 474	J6	Rocky Mts.: Denali 6,190	K6
ea 2,274	E7	Lake Baikal 31	F7	Ob-Irtysh 5,410	G7	Milwaukee Deep 9,200	H7	Honshu 231	J7	East Africa: Mt. Kilimanjaro 5,895	K7
an Sea 1,942	E8	Great Bear Lake 31	F8	Congo 4,670	G8	Bouganville Trench 9,140	H8	Great Britain 230	J8	Caucasus: Elbrus 5,640	K8
Mexico 1,813	E9	Lake Malawi 30	F9	Mekong 4,500	G9	Aleutian Trench 7,822	H9	Victoria Island 212	J9	Antarctica: Vinson Massif 4,897	K9
Okhotsk 1,528	E10	Great Slave Lake 29	F10	Amur 4,442	G10	South Sandwich Island Trench 7,235	H10	Ellesmere Island 197	J10	Alps: Mt. Blanc 4,808	K10

COPYRIGHT PHILIP'S

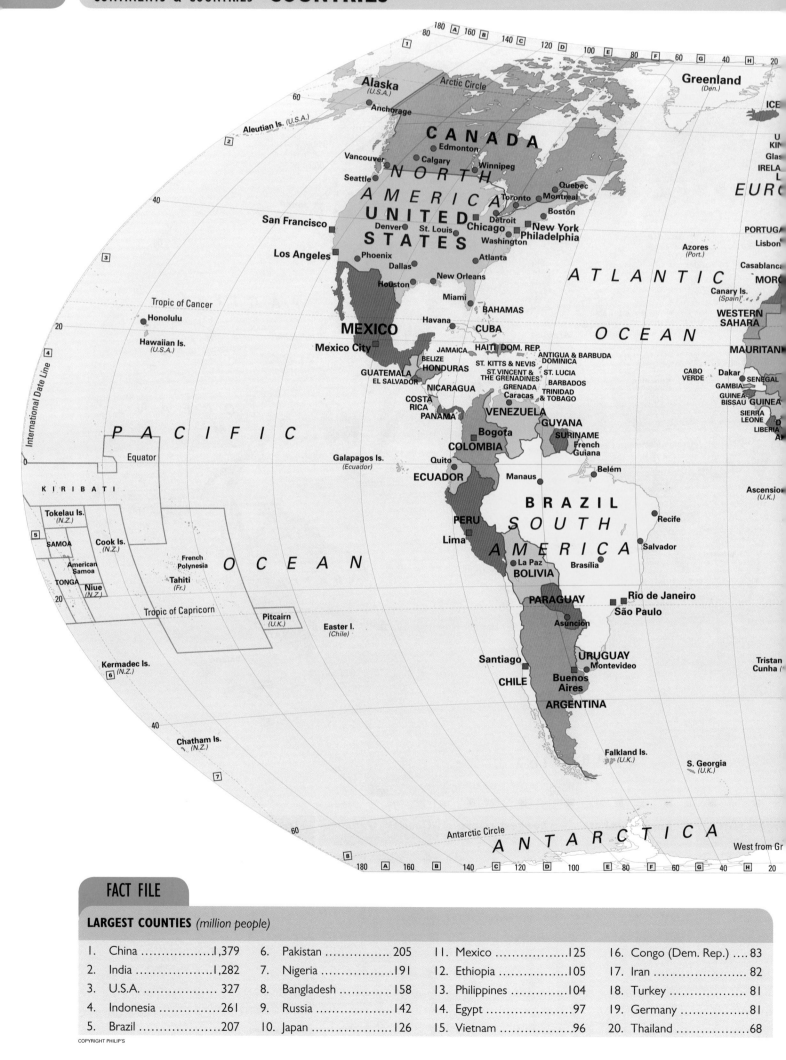

FACT FILE

LARGEST COUNTIES *(million people)*

1. China	1,379	6. Pakistan	205	11. Mexico	125	16. Congo (Dem. Rep.)	83
2. India	1,282	7. Nigeria	191	12. Ethiopia	105	17. Iran	82
3. U.S.A.	327	8. Bangladesh	158	13. Philippines	104	18. Turkey	81
4. Indonesia	261	9. Russia	142	14. Egypt	97	19. Germany	81
5. Brazil	207	10. Japan	126	15. Vietnam	96	20. Thailand	68

Equatorial Scale 1:100 000 000
This distance is 4,000 kilometres

SMALLEST COUNTRIES *(thousand people)*

1. Vatican City 1	6. San Marino 34	11. Andorra 86	16. Tonga 106
2. Nauru 10	7. Liechtenstein 38	12. Seychelles 94	17. Kiribati 108
3. Tuvalu 11	8. St. Kitts & Nevis 52	13. Antigua & Barbuda 95	18. Grenada 112
4. Palau 21	9. Dominica 74	14. St. Vincent & The Grenadines ...102	19. St. Lucia 164
5. Monaco 31	10. Marshall Islands 75	15. Micronesia, Fed. States of 104	20. Samoa 200

COPYRIGHT PHILIP'S

AT A GLANCE

- North America is the third largest continent. It is half the size of Asia. It stretches almost from the Equator to the North Pole.
- Three countries – Canada, the United States and Mexico – make up most of the continent.
- Greenland, the largest island in the world, is part of North America.

- In the east there are several large lakes. These are called the Great Lakes. A large waterfall called Niagara Falls is between Lake Erie and Lake Ontario. The St Lawrence River connects the Great Lakes with the Atlantic Ocean.
- North and South America are joined by the Isthmus of Panama.

Mount Denali (once named McKinley) is the highest peak of the Rocky mountain chain.

Lake Ontario, one of the Great Lakes, flows over the Niagara Falls into Lake Erie.

Height of the land (metres)

Over 4000
2000–4000
1000–2000
400–1000
200–400
Sea level   0–100
Below sea level

Scale 1:63 000 000
This distance is 2000 kilometres

COPYRIGHT PHILIP'S

FACT FILE

**LARGEST COUNTRIES:
BY AREA**
(thousand square kilometres)

1.	Canada	9,971
2.	United States	9,629
3.	Mexico	1,958
4.	Nicaragua	129
5.	Honduras	112
6.	Cuba	111
7.	Guatemala	109
8.	Panama	76
9.	Costa Rica	51
10.	Dominican Republic	49

**LARGEST COUNTRIES:
BY POPULATION**
(million people)

1.	United States	327
2.	Mexico	125
3.	Canada	36
4.	Guatemala	15
5.	Cuba	11
6.	Haiti	11
7.	Dominican Republic	10
8.	Honduras	9
9.	El Salvador	6
10.	Nicaragua	6

LARGEST CITIES
(million people)

1.	Mexico City, Mexico	21.2
2.	New York, USA	20.1
3.	Los Angeles, USA	13.3
4.	Chicago, USA	9.6
5.	Dallas-Fort Worth, USA	7.0
6.	Houston, USA	6.5
7.	Toronto, Canada	6.1
8.	Philadelphia, USA	6.1
9.	Washington, DC, USA	6.0
10.	Miami, USA	5.9

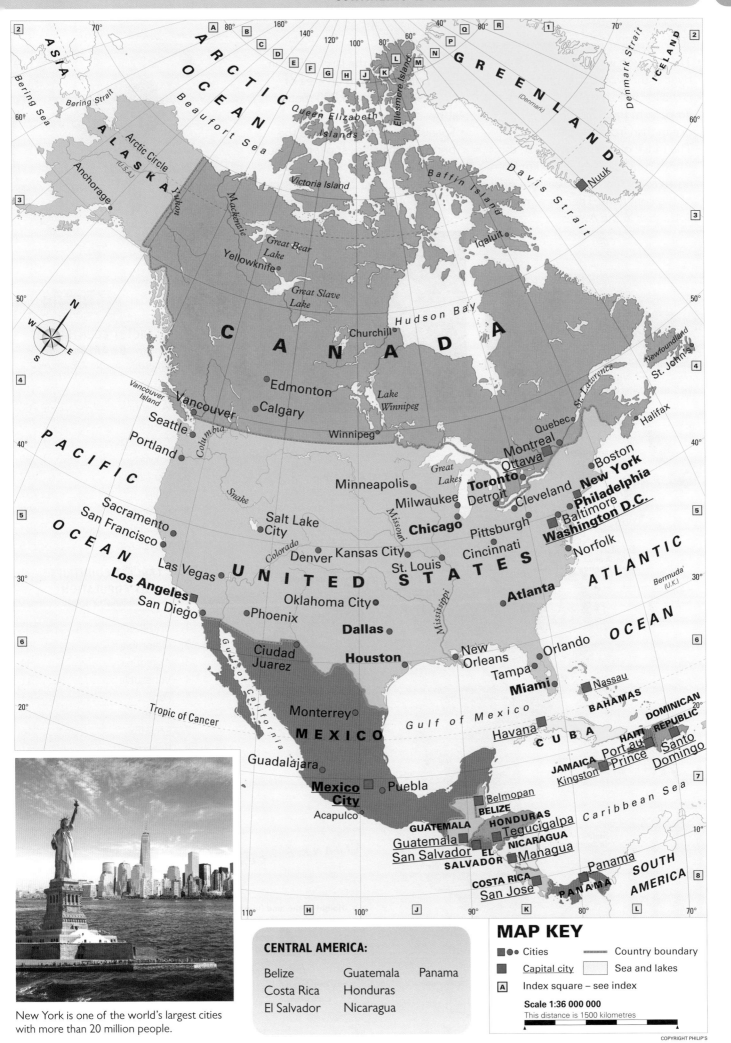

ASIA

ARCTIC OCEAN

Bering Sea
Bering Strait
Beaufort Sea

GREENLAND
(Denmark)

Denmark Strait

ICELAND

ALASKA
(USA)

Arctic Circle

Anchorage

Yukon

Mackenzie

Queen Elizabeth
Islands

Ellesmere Island

Nuuk

Davis Strait

Victoria Island

Great Bear
Lake

Yellowknife

Baffin Island

Iqaluit

C A N A D A

Great Slave
Lake

Hudson Bay

Churchill

Newfoundland

St. John's

Lake
Winnipeg

St. Lawrence

Halifax

Vancouver
Island

Edmonton

Vancouver

Calgary

Columbia

Seattle

Winnipeg

Quebec

Montreal

Boston

Portland

PACIFIC

Ottawa

Minneapolis

Toronto

New York

Milwaukee Detroit Cleveland

Philadelphia

Snake

Chicago

Pittsburgh

Baltimore

Washington D.C.

Sacramento

Salt Lake
City

Denver

Kansas City

Cincinnati

Norfolk

San Francisco

Colorado

Missouri

St. Louis

U N I T E D   S T A T E S

ATLANTIC

OCEAN

Las Vegas

Los Angeles

San Diego

Oklahoma City

Atlanta

Bermuda
(U.K.)

Phoenix

Dallas

Mississippi

OCEAN

Gulf of California

Houston

New
Orleans

Orlando

Ciudad
Juarez

Tampa

Miami

Nassau

Monterrey

Tropic of Cancer

Gulf of Mexico

BAHAMAS

DOMINICAN
REPUBLIC

M E X I C O

Havana

CUBA

HAITI

Santo

Guadalajara

JAMAICA

Port au
Prince

Domingo

Mexico
City

Puebla

Kingston

Belmopan

Caribbean Sea

Acapulco

BELIZE

GUATEMALA

HONDURAS

Tegucigalpa

SOUTH

Guatemala

NICARAGUA

San Salvador

EL
SALVADOR

Managua

Panama

AMERICA

COSTA RICA

PANAMA

San Jose

New York is one of the world's largest cities
with more than 20 million people.

CENTRAL AMERICA:

Belize   Guatemala   Panama

Costa Rica   Honduras

El Salvador   Nicaragua

MAP KEY

■●● Cities   ——— Country boundary

■ Capital city   Sea and lakes

Ⓐ Index square – see index

Scale 1:36 000 000
This distance is 1500 kilometres

COPYRIGHT PHILIP'S

AT A GLANCE

- The Amazon is the second longest river in the world. The Nile in Africa is the longest river, but more water flows from the Amazon into the ocean than from any other river.
- The range of mountains called the Andes runs for over 7,500 km from north to south on the western side of the continent. There are many volcanoes in the Andes.
- Lake Titicaca is the largest lake in the continent. It has an area of 8,300 sq km and is 3,800 metres above sea level.
- Spanish and Portuguese are the principal languages spoken in South America.
- Brazil is the largest country in area and population, and the largest city, Sao Paulo.

Snow-covered Andes Mountains tower above Lake Titicaca which is 3812 metres high.

The Amazon river system empties into the south Atlantic Ocean at the Equator.

Scale 1:60 000 000
This distance is 2000 kilometres

Height of the land (metres)

Over 4000	
2000–4000	
1000–2000	
400–1000	
200–400	
Sea level	0–100
	Below sea level

COPYRIGHT PHILIP'S

FACT FILE

LARGEST COUNTRIES: BY AREA
(thousand square kilometres)

1. Brazil 8,514
2. Argentina. 2,780
3. Peru. 1,285
4. Colombia 1,139
5. Bolivia. 1,099
6. Venezuela 912
7. Chile 757
8. Paraguay 407
9. Ecuador. 284
10. Guyana 215

LARGEST COUNTRIES: BY POPULATION
(million people)

1. Brazil 207
2. Colombia 48
3. Argentina 44
4. Venezuela 31
5. Peru 31
6. Chile 18
7. Ecuador. 16
8. Bolivia 11
9. Paraguay 7
10. Uruguay. 3

LARGEST CITIES
(million people)

1. Sao Paulo, Brazil 21.3
2. Buenos Aires, Argentina 15.3
3. Rio de Janeiro, Brazil . . . 13.0
4. Lima, Peru 10.1
5. Bogota, Colombia 10.0
6. Santiago, Chile 6.5
7. Belo Horizonte, Brazil . . . 5.8
8. Brasilia, Brazil 4.2
9. Medellin, Colombia 4.0
10. Fortaleza, Brazil. 3.9

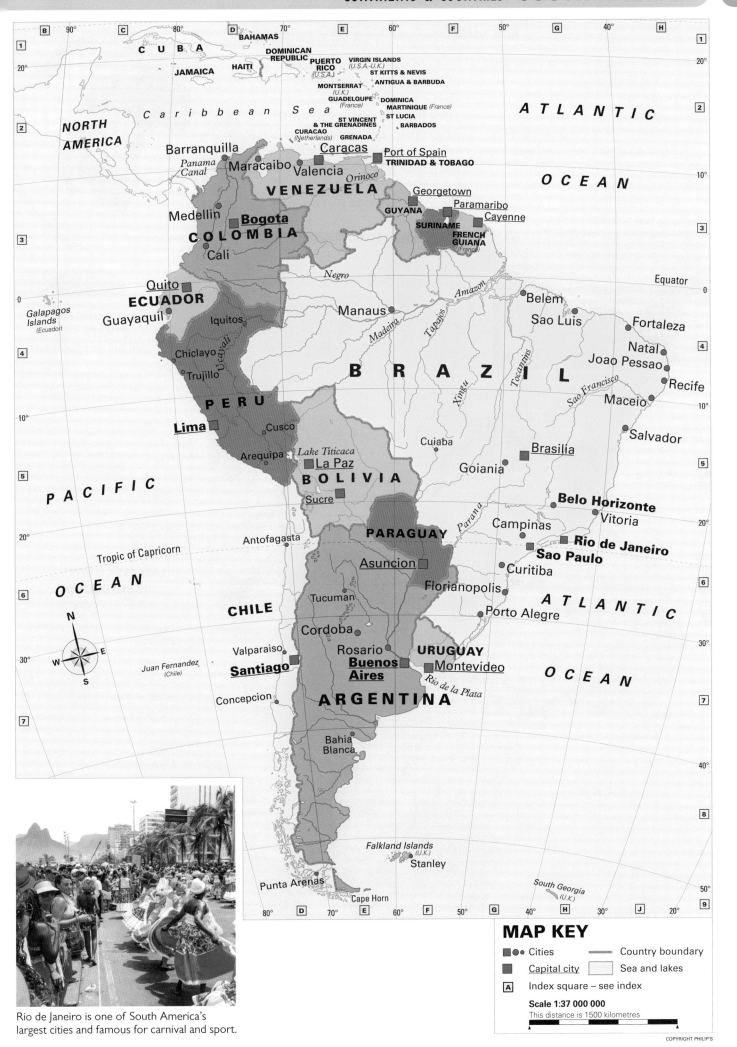

B 90° C 80° D 70° E 60° F 50° G 40° H
1 | 1

CUBA BAHAMAS 20°

JAMAICA HAITI DOMINICAN REPUBLIC

PUERTO RICO (U.S.A.) VIRGIN ISLANDS (U.S.A.-U.K.)
ST KITTS & NEVIS
ANTIGUA & BARBUDA

MONTSERRAT (U.K.)
GUADELOUPE (France) DOMINICA
MARTINIQUE (France)
ST LUCIA
ST VINCENT & THE GRENADINES BARBADOS
CURACAO (Netherlands) GRENADA

ATLANTIC 2

NORTH AMERICA

Caribbean Sea

Barranquilla
Panama Canal
Maracaibo
Valencia
Caracas
Port of Spain
TRINIDAD & TOBAGO

OCEAN 10°

Orinoco

VENEZUELA

Georgetown
Paramaribo
Cayenne

GUYANA
SURINAME
FRENCH GUIANA (France)

3

Medellin
Bogota
COLOMBIA
Cali

Negro
Amazon

Equator 0

Quito
ECUADOR
Guayaquil
Iquitos

Belem
Sao Luis
Fortaleza
Natal
Joao Pessao
Recife

4

Galapagos Islands (Ecuador)

Chiclayo
Trujillo
Ucayali

B R A Z I L

Madeira
Tapajos
Xingu
Tocantins
Sao Francisco

Maceio

PERU

10°

Lima
Cusco
Arequipa
Lake Titicaca
La Paz
BOLIVIA
Sucre

Cuiaba

Brasilia
Goiania

Salvador

5

PACIFIC

Belo Horizonte
Vitoria

20°

Antofagasta
Parana
PARAGUAY
Asuncion
Campinas
Sao Paulo
Curitiba

Rio de Janeiro

Tropic of Capricorn

OCEAN

6

Tucuman

Florianopolis
Porto Alegre

ATLANTIC

CHILE

Cordoba

Valparaiso
Juan Fernandez (Chile)
Rosario
Santiago
Buenos Aires
Montevideo
URUGUAY

Rio de la Plata

OCEAN

30°

Concepcion
ARGENTINA

7

Bahia Blanca

40°

8

Falkland Islands (U.K.)
Stanley

Punta Arenas
Cape Horn
South Georgia (U.K.)

50°

9

80° D 70° E 60° F 50° G 40° H 30° J 20°

N W E S

Rio de Janeiro is one of South America's largest cities and famous for carnival and sport.

MAP KEY

▪●● Cities — — — Country boundary

▪ <u>Capital city</u> □ Sea and lakes

A Index square – see index

Scale 1:37 000 000
This distance is 1500 kilometres

COPYRIGHT PHILIP'S

AT A GLANCE

- Africa is the second largest continent. Asia is the largest.
- There are over 50 countries, some of them small in population. The population of Africa is growing more quickly than any other continent.
- Parts of Africa have a dry, desert climate. Some other parts are tropical.
- The highest mountains run from north to south on the eastern side of Africa.
- The Great Rift Valley is a volcanic valley that was formed 10 to 20 million years ago through long splits in the Earth's crust. Mount Kenya and Kilimanjaro are old volcanoes in the Rift Valley area.
- The Sahara is the largest desert in the world.

Sand dunes of the Sahara result from the dry, desert climate of much of North Africa

Lake Tanganyika is in a section of the Great Rift Valley. The entire valley is 6000 km long.

FACT FILE

LARGEST COUNTRIES: BY AREA
(thousand square kilometres)

1.	Algeria	2,382
2.	Dem. Rep. of the Congo	2,345
3.	Sudan	1,886
4.	Libya	1,759
5.	Chad	1,284
6.	Niger	1,267
7.	Angola	1,247
8.	Mali	1,240
9.	South Africa	1,221
10.	Ethiopia	1,104

LARGEST COUNTRIES: BY POPULATION
(million people)

1.	Nigeria	191
2.	Ethiopia	105
3.	Egypt	97
4.	Dem. Rep. of the Congo	83
5.	South Africa	55
6.	Tanzania	54
7.	Kenya	48
8.	Algeria	41
9.	Uganda	40
10.	Sudan	37

LARGEST CITIES
(million people)

1.	Cairo, Egypt	17.1
2.	Lagos, Nigeria	13.7
3.	Kinshasa, Dem. Rep. of the Congo	12.1
4.	Johannesburg, S. Africa	9.6
5.	Luanda, Angola	5.7
6.	Dar es Salaam, Tanzania	5.4
7.	Khartoum, Sudan	5.2
8.	Abidjan, Côte d'Ivoire	5.0
9.	Alexandria, Egypt	4.9
10.	Nairobi, Kenya	4.0

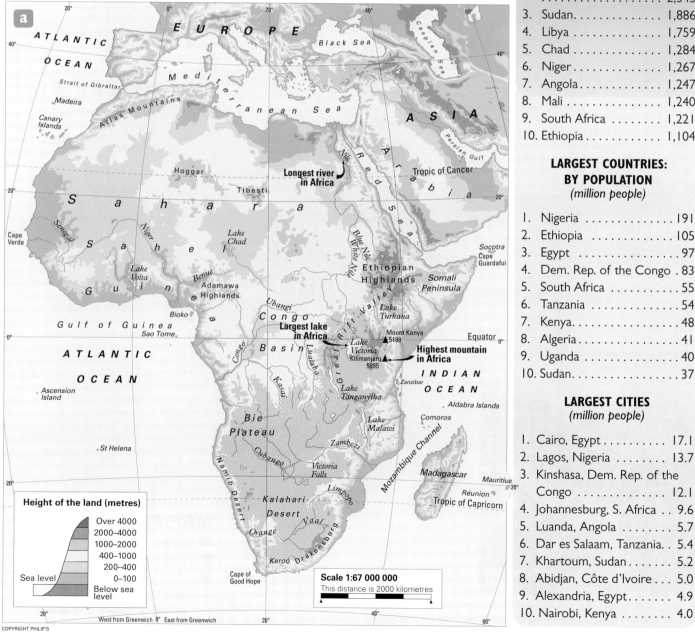

Height of the land (metres)

Over 4000
2000–4000
1000–2000
400–1000
200–400
0–100
Sea level
Below sea level

Scale 1:67 000 000
This distance is 2000 kilometres

COPYRIGHT PHILIP'S

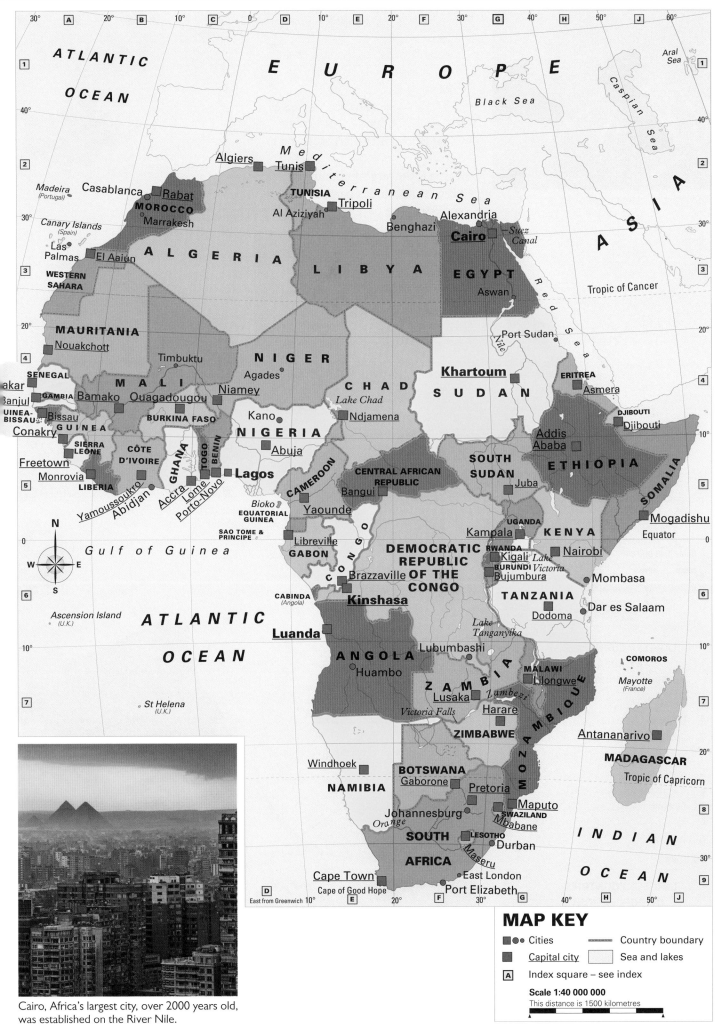

ATLANTIC OCEAN

EUROPE

ASIA

Aral Sea

Black Sea

Caspian Sea

Mediterranean Sea

Madeira (Portugal)

Casablanca

Algiers
Tunis
TUNISIA
Tripoli
Al Aziziyah

Rabat
MOROCCO
Marrakesh

Canary Islands (Spain)

Las Palmas
El Aaiun
WESTERN SAHARA

ALGERIA

LIBYA

Benghazi

Alexandria
Cairo
Suez Canal

EGYPT
Aswan

Tropic of Cancer

MAURITANIA
Nouakchott

NIGER

Timbuktu
Agades

Khartoum

Port Sudan

Nile

Red Sea

ERITREA
Asmera

SUDAN

DJIBOUTI
Djibouti

SENEGAL
akar
GAMBIA
Banjul
UINEA-BISSAU
Bissau
Conakry
GUINEA
SIERRA LEONE
Freetown
Monrovia
LIBERIA

MALI
Bamako
Ouagadougou
BURKINA FASO

Niamey
Kano

CHAD
Lake Chad
Ndjamena

NIGERIA
Abuja
Lagos

CÔTE D'IVOIRE
GHANA
TOGO
BENIN
Accra
Lome
Porto-Novo

Yamoussoukro
Abidjan

Bioko
EQUATORIAL GUINEA
SAO TOME & PRINCIPE

CAMEROON
Yaounde

CENTRAL AFRICAN REPUBLIC
Bangui

SOUTH SUDAN
Juba

Addis Ababa
ETHIOPIA

SOMALIA
Mogadishu

Equator

Libreville
GABON
CONGO

DEMOCRATIC REPUBLIC OF THE CONGO

UGANDA
Kampala
RWANDA
Kigali
BURUNDI
Bujumbura
Lake Victoria

KENYA
Nairobi
Mombasa

Gulf of Guinea

Ascension Island (U.K.)

ATLANTIC OCEAN

Brazzaville
CABINDA (Angola)
Kinshasa

Luanda

St Helena (U.K.)

ANGOLA
Huambo

Lubumbashi

Lake Tanganyika

TANZANIA
Dodoma
Dar es Salaam

ZAMBIA
Lusaka
Zambezi
Victoria Falls

MALAWI
Lilongwe

COMOROS
Mayotte (France)

Windhoek

Harare
ZIMBABWE

MOZAMBIQUE

Antananarivo
MADAGASCAR
Tropic of Capricorn

NAMIBIA

BOTSWANA
Gaborone

Johannesburg
Orange
SOUTH AFRICA

Pretoria
Maputo
SWAZILAND
Mbabane
LESOTHO
Maseru
Durban

INDIAN OCEAN

Cape Town
Cape of Good Hope
East from Greenwich
East London
Port Elizabeth

Cairo, Africa's largest city, over 2000 years old, was established on the River Nile.

MAP KEY

■●● Cities	〰〰 Country boundary
■ Capital city	☐ Sea and lakes
A Index square – see index	

Scale 1:40 000 000
This distance is 1500 kilometres

COPYRIGHT PHILIP'S

The Ural Mountain range divides one land mass into the continents of Europe and Asia.

AT A GLANCE

- Europe is one-fifth the size of Asia. Australia is slightly smaller than Europe.
- The Ural Mountains are viewed as the eastern boundary of Europe.
- Great Britain is the largest island in Europe
- Russia is the largest country in Europe. It includes parts of Europe and Asia. The part in Asia is far larger.

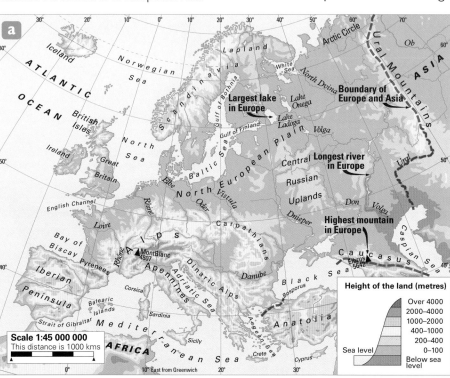

Height of the land (metres)

	Over 4000
	2000–4000
	1000–2000
	400–1000
	200–400
Sea level	0–100
	Below sea level

Scale 1:45 000 000
This distance is 1000 kms

FACT FILE

LARGEST COUNTRIES: BY AREA
(thousand square kilometres)

1.	Russia	17,075
2.	Ukraine	604
3.	France	551
4.	Spain	497
5.	Sweden	450
6.	Germany	357
7.	Finland	338
8.	Norway	324
9.	Poland	323
10.	Italy	301

LARGEST COUNTRIES: BY POPULATION
(million people)

1.	Russia	142
2.	Germany	81
3.	France	67
4.	United Kingdom	65
5.	Italy	62

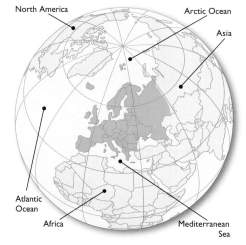

The Highlands of Scotland lie in the northern part of Great Britain. This is Europe's largest island.

Moscow, the largest city in Europe, is capital of Russia. Red Square is the heart of the city.

ARCTIC OCEAN

North Cape

Lofoten Islands

Narvik

Murmansk

N O R W A Y

S W E D E N

F I N L A N D

White Sea

Arkhangelsk

North Devina

Trondheim

Gulf of Bothnia

Lake Onega

Lake Ladoga

etland ands (U.K.)

Bergen

Oslo

Helsinki

Stockholm

Gulf of Finland

St. Petersburg

Volga

R U S S I A

Ufa

Gothenburg

Tallinn

ESTONIA

Kazan

Samara

North Sea

DENMARK

Riga

LATVIA

Moscow

Copenhagen

LITHUANIA

Baltic Sea

Gdansk

KALININGRAD (Russia)

Vilnius

Minsk

BELARUS

Voronezh

Ural

Hamburg

ERLANDS

Amsterdam

The Hague

sels

Berlin

Warsaw

Elbe

Vistula

Oder

Don

Volgograd

Volga

GERMANY

P O L A N D

GIUM

Frankfurt

Krakow

Kiev

Kharkov

LUXEMBOURG

uxembourg

Prague

CZECHIA

Rhine

Munich

U K R A I N E

Lvov

Dnepropetrovsk

Donetsk

Rostov

Caspian Sea

Bern

Vienna

Bratislava

LIECHTENSTEIN

A U S T R I A

Budapest

MOLDOVA

Kishinev

Odessa

Sea of Azov

Krasnodar

SWITZERLAND

os

Ljubljana

SLOVENIA

HUNGARY

Zagreb

C R O A T I A

R O M A N I A

Crimea

urin

Milan

BOSNIA-HERZEGOVINA

Belgrade

Sevastopol

MONACO

SAN MARINO

I T A L Y

Adriatic Sea

Sarajevo

SERBIA

Bucharest

Danube

Black Sea

eilles

rsica ance)

Podgorica

MONTENEGRO

Pristina

KOSOVO

BULGARIA

Sofia

Rome

Skopje

MACEDONIA

Istanbul

Naples

Tirane

ALBANIA

Ankara

T U R K E Y

Palermo

Sicily

G R E E C E

Aegean Sea

Izmir

A S I A

rranean

Athens

Valletta

MALTA

Sea

Crete

Nicosia

CYPRUS

A S I A

SLOVAKIA

MAP KEY

- Cities
- Capital city
- Index square – see index
- Country boundary
- Sea and lakes

Scale 1:20 000 000
This distance is 750 kilometres

COPYRIGHT PHILIP'S

Mount Everest viewed from Nepal.

AT A GLANCE

- Asia is the largest continent. It is twice the size of North America.
- It is a continent of long rivers. Many of Asia's rivers are longer than Europe's longest river.
- Asia contains more than half of the world's population.
- Mount Everest, 8850m, is the world's highest peak. It lies on the border between Nepal and China.

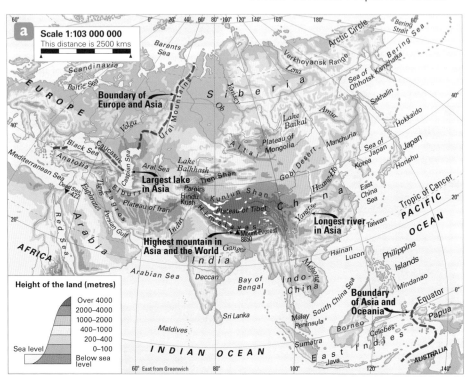

Scale 1:103 000 000
This distance is 2500 kms

Height of the land (metres)
- Over 4000
- 2000–4000
- 1000–2000
- 400–1000
- 200–400
- 0–100
- Sea level
- Below sea level

FACT FILE

LARGEST COUNTRIES: BY AREA
(thousand square kilometres)

1. China 9,597
2. India 3,287
3. Kazakhstan 2,725
4. Saudi Arabia 2,150
5. Indonesia 1,905
6. Iran 1,648
7. Mongolia 1,566
8. Pakistan 796
9. Turkey 775
10. Myanmar 677

LARGEST COUNTRIES: BY POPULATION
(million people)

1. China 1,379
2. India 1,282
3. Indonesia 261
4. Pakistan 205
5. Bangladesh 158

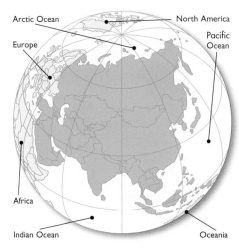

The Yangtse river winds 6000 km across China from west to east.

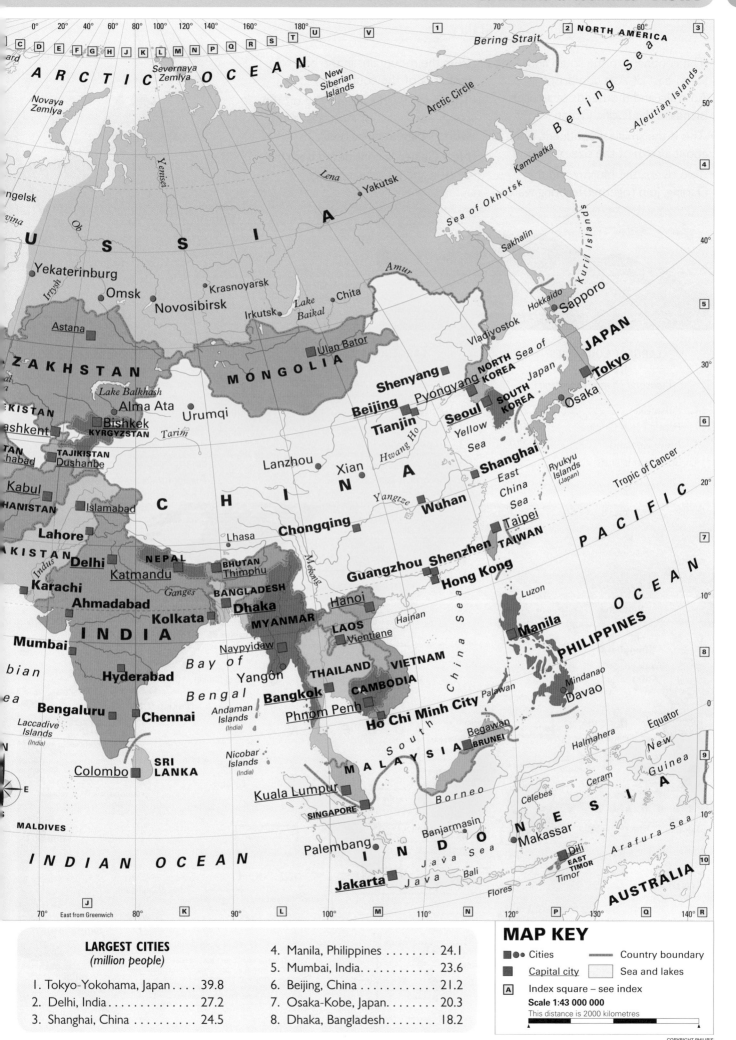

ARCTIC OCEAN
Severnaya Zemlya
Novaya Zemlya
New Siberian Islands
Arctic Circle
Bering Strait
NORTH AMERICA
Bering Sea
Aleutian Islands

RUSSIA
ngelsk
vina
Yenisei
Ob
Irtysh
Lena
Amur
Yakutsk
Krasnoyarsk
Chita
Omsk
Novosibirsk
Irkutsk
Lake Baikal
Yekaterinburg
Sea of Okhotsk
Kamchatka
Kuril Islands
Sakhalin
Vladivostok
Hokkaido
Sapporo
JAPAN
Tokyo

KAZAKHSTAN
Astana
Lake Balkhash
Alma Ata
Urumqi
Tarim
MONGOLIA
Ulan Bator
Shenyang
Beijing
Tianjin
NORTH KOREA
Pyongyang
Sea of Japan
SOUTH KOREA
Seoul
Osaka
Yellow Sea
Shanghai
East China Sea
Ryukyu Islands (Japan)
Tropic of Cancer

KIKISTAN
ashkent
Bishkek
KYRGYZSTAN
TAN
habad
TAJIKISTAN
Dushanbe
Lanzhou
Xian
CHINA
Hwang Ho
Yangtze
Wuhan
Chongqing
Taipei
TAIWAN

Kabul
HANISTAN
Islamabad
Lhasa
NEPAL
Katmandu
BHUTAN
Thimphu
Mekong
Guangzhou **Shenzhen**
Hong Kong
PACIFIC

Lahore
AKISTAN
Indus
Delhi
Ganges
BANGLADESH
Dhaka
Karachi
Ahmadabad
Kolkata
INDIA
MYANMAR
Hanoi
LAOS
Vientiane
Hainan
Luzon
Manila
PHILIPPINES
OCEAN

Mumbai
bian
ea
Naypyidaw
Bay of Bengal
Hyderabad
Yangon
THAILAND
VIETNAM
Mindanao
Davao

Bengaluru
Chennai
Andaman Islands (India)
Bangkok
CAMBODIA
Phnom Penh
Ho Chi Minh City
Palawan
South China Sea
Equator
Halmahera
New Guinea

Laccadive Islands (India)
Colombo
SRI LANKA
Nicobar Islands (India)
Begawan
BRUNEI
Celebes
Ceram

MALDIVES
MALAYSIA
Borneo
INDONESIA
Makassar
Arafura Sea

INDIAN OCEAN
Kuala Lumpur
SINGAPORE
Banjarmasin
Palembang
Java Sea
Dili
EAST TIMOR
Timor

Jakarta
Java
Bali
Flores
AUSTRALIA

East from Greenwich

LARGEST CITIES
(million people)

1. Tokyo-Yokohama, Japan 39.8	4. Manila, Philippines 24.1
2. Delhi, India 27.2	5. Mumbai, India 23.6
3. Shanghai, China 24.5	6. Beijing, China 21.2
	7. Osaka-Kobe, Japan 20.3
	8. Dhaka, Bangladesh 18.2

MAP KEY

●● Cities — Country boundary
■ Capital city ☐ Sea and lakes
Ⓐ Index square – see index
Scale 1:43 000 000
This distance is 2000 kilometres

COPYRIGHT PHILIP'S

AT A GLANCE

- The continent to the south and south-east of Asia comprises Australia and thousands of smaller islands in the Pacific Ocean
- This is the smallest continent, only about a sixth of the size of Asia.
- Two rivers, the Murray and the Darling, join together to bring water from the mountains in the east to a vast area of the country.
- The Great Barrier Reef, lying off the north-east coast of Australia, is the world's biggest coral reef.
- New Guinea is considered both part of Asia and a Pacific country.

Many Pacific countries have islands which are active volcanoes, such as this one in Vanuatu.

FACT FILE

LARGEST COUNTRIES: BY AREA (thousand square kilometres)		LARGEST COUNTRIES: BY POPULATION (million people)		LARGEST CITIES (million people)	
1. Australia	7,741	1. Australia	23	1. Sydney, Australia	4.5
2. Papua New Guinea	463	2. Papua New Guinea	7	2. Melbourne, Australia	4.3
3. New Zealand	270	3. New Zealand	5	3. Brisbane, Australia	2.2
4. Solomon Islands	26	4. Fiji	0.9	4. Perth, Australia	1.9
5. Fiji	18	5. Solomon Islands	0.6	5. Auckland, New Zealand	1.4

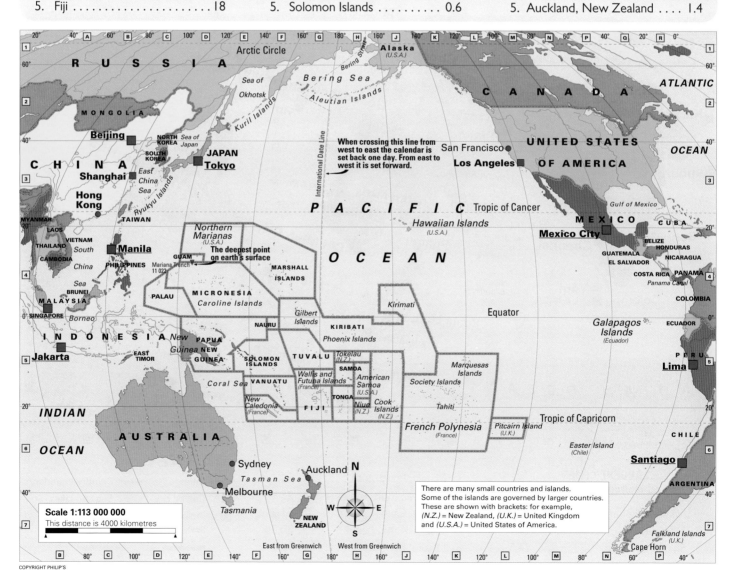

When crossing this line from west to east the calendar is set back one day. From east to west it is set forward.

The deepest point on earth's surface

Mariana Trench 11 022

There are many small countries and islands. Some of the islands are governed by larger countries. These are shown with brackets: for example, (N.Z.) = New Zealand, (U.K.) = United Kingdom and (U.S.A.) = United States of America.

Scale 1:113 000 000
This distance is 4000 kilometres

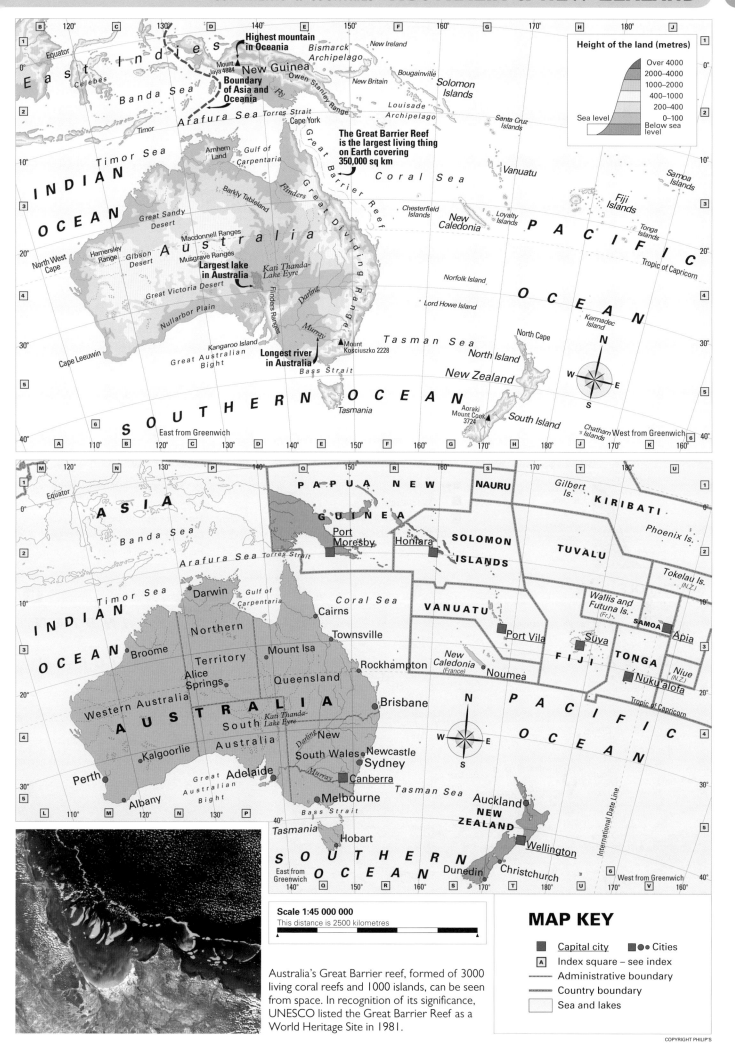

Height of the land (metres)

	Over 4000
	2000–4000
	1000–2000
	400–1000
	200–400
Sea level	0–100
	Below sea level

Top map labels:

East Indies
Equator
Celebes
Timor
Banda Sea
INDIAN OCEAN
Timor Sea
Arafura Sea
Torres Strait
Cape York
Arnhem Land
Gulf of Carpentaria
Barkly Tableland
Great Sandy Desert
Hamersley Range
North West Cape
Gibson Desert
Macdonnell Ranges
Musgrave Ranges
Great Victoria Desert
Nullarbor Plain
Flinders Ranges
Kati Thanda-Lake Eyre
Darling
Murray
Cape Leeuwin
Great Australian Bight
Kangaroo Island
Bass Strait
Tasmania
SOUTHERN OCEAN
East from Greenwich

Highest mountain in Oceania
Mount Jaya 4884
Boundary of Asia and Oceania
New Guinea
Owen Stanley Range
Fly
Bismarck Archipelago
New Ireland
New Britain
Bougainville
Solomon Islands
Louisade Archipelago
The Great Barrier Reef is the largest living thing on Earth covering 350,000 sq km
Coral Sea
Great Dividing Range
Great Barrier Reef
Chesterfield Islands
New Caledonia
Loyalty Islands
Santa Cruz Islands
Vanuatu
Fiji Islands
Samoa Islands
Tonga Islands
PACIFIC OCEAN
Tropic of Capricorn
Norfolk Island
Lord Howe Island
Kermadec Island
Largest lake in Australia
Australia
Longest river in Australia
Mount Kosciuszko 2228
Tasman Sea
North Cape
North Island
New Zealand
Aoraki Mount Cook 3724
South Island
Chatham Islands
West from Greenwich

Bottom map labels:

Equator
ASIA
Banda Sea
Arafura Sea
Torres Strait
INDIAN OCEAN
Timor Sea
Darwin
Gulf of Carpentaria
Broome
Northern Territory
Alice Springs
Western Australia
AUSTRALIA
Kalgoorlie
Perth
Kati Thanda-Lake Eyre
South Australia
Australia
Great Australian Bight
Albany
Adelaide
Cairns
Townsville
Mount Isa
Rockhampton
Queensland
Brisbane
Darling
New South Wales
Newcastle
Sydney
Canberra
Murray
Melbourne
Bass Strait
Tasmania
Hobart
SOUTHERN OCEAN
East from Greenwich

PAPUA NEW GUINEA
Port Moresby
Honiara
SOLOMON ISLANDS
NAURU
Gilbert Is.
KIRIBATI
Phoenix Is.
TUVALU
Tokelau Is. (N.Z.)
Wallis and Futuna Is. (Fr.)
Coral Sea
VANUATU
Port Vila
New Caledonia (France)
Noumea
Suva
FIJI
SAMOA
Apia
TONGA
Nuku'alofa
Niue (N.Z.)
PACIFIC OCEAN
Tropic of Capricorn
Tasman Sea
Auckland
NEW ZEALAND
Wellington
Christchurch
Dunedin
International Date Line
West from Greenwich

Scale 1:45 000 000
This distance is 2500 kilometres

Australia's Great Barrier reef, formed of 3000 living coral reefs and 1000 islands, can be seen from space. In recognition of its significance, UNESCO listed the Great Barrier Reef as a World Heritage Site in 1981.

MAP KEY

■ Capital city	■●● Cities
Ⓐ Index square – see index	
——— Administrative boundary	
≈≈≈ Country boundary	
▢ Sea and lakes	

COPYRIGHT PHILIP'S

NORTH POLE

PACIFIC OCEAN

Anchorage

Nizhne Kolymsk

Wrangel I. (Russia)
Tiksi
New Siberian Is. (Russia)
Laptev Sea
ASIA
Novosibirsk

Cape Barrow
ARCTIC OCEAN
Taimyr Peninsula
Norilsk

Beaufort Sea
C. Chelyuskin
Severnaya Zemlya (Russia)

Inuvik
Vorkuta

Vancouver
Mackenzie
Banks Island (Canada)
Queen Elizabeth Islands (Canada)
North Magnetic Pole
North Pole
Franz Josef Land (Russia)
Novaya Zemlya (Russia)
Kara Sea
Ural Mts.

Great Bear L.
Barents Sea

Edmonton
Yellowknife
Great Slave L.
Victoria Island (Canada)
C. Morris Jesup
Arkhangelsk
Murmansk
N. Dvina

NORTH AMERICA
Ellesmere Island (Canada)
Thule
Svalbard (Norway)
North Cape
Tromsø

Churchill
Nelson
Baffin Island (Canada)
Baffin Bay
80°N
Greenland
Greenland Sea
Moscow
EUROPE

Hudson Bay
Jan Mayen I. (Norway)
Arctic Circle
Scandinavia
St. Petersburg
Baltic Sea

L. Michigan
L. Superior
Iqaluit
Davis Strait
70°N
Denmark Strait
Iceland
Faroe Is. (Denmark)
Oslo
Black Sea

Chicago
L. Huron
Nuuk (Godthåb)
Reykjavik
North Sea
Edinburgh

C. Farewell
60°N
British Isles
West from Greenwich
0° East from Greenwich

Scale 1:50 000 000
This distance is 2500 kilometres

Height of the land (metres)
Over 4000
2000–4000
1000–2000
400–1000
200–400
Sea level
0–200
Below sea level

Cities

Capital city

Index square – see index

Davis (Austr.) Research station and the country which runs it.

Limit of permanently frozen sea

Icebergs

Furthest extent of icebergs

Land permanently covered with ice

Height of ice (in metres)

SOUTH POLE

ATLANTIC OCEAN
South Sandwich Trench
▼ 8265
South Sandwich Is. (U.K.)
Syowa (Japan)
Enderby Land
Mawson (Austr.)
C. Darnley
INDIAN OCEAN

South Georgia (U.K.)
Antarctic Circle
Maltri (India)
Prince Charles Mts
Davis (Austr.)

Sanae (S. Africa)
Queen Maud Land
American Highland

South Orkney Is.
Halley (U.K.)
Coats Land
80°S

Scotia Sea
Weddell Sea
ANTARCTICA
Casey (Austr.)

South Shetland Is.
O'Higgins (Chile)
Esperanza (Arg.)
Berkner I.
Ronne Ice Shelf
Vostok (Russia)
Wilkes Land

Falkland Is. (U.K.)
Palmer (U.S.A.)
Rothera (U.K.)
Antarctic Peninsula
South Pole
Amundsen-Scott (U.S.A.)
Beardmore Glacier
Queen Maud Ra.

C. Horn
Alexander I.
Bellingshausen Sea
Ellsworth Land
Vinson Massif 5140
▲ Mt. Markham 4349
South Magnetic Pole

Punta Arenas
Tierra del Fuego
Drake Passage
Marie Byrd Land
Ross Ice Shelf
Scott (N.Z.)
McMurdo (U.S.A.)
Adélie Land
Dumont d'Urville (France)

SOUTH AMERICA
Strait of Magellan
Victoria Land
Ross Sea

C. Adare
Balleny Is.

Tasmania
Hobart
AUSTRALIA

Antarctic Circle
SOUTHERN OCEAN
Macquarie I. (Australia)
Campbell I. (N.Z.)
Auckland I.
West from Greenwich
East from Greenwich

Scale 1:50 000 000
This distance is 2500 kilometres

COPYRIGHT PHILIP'S

Country	Flag	Capital	Area (sq km)	Population (thousands)
NORTH AMERICA				
TIGUA & RBUDA		St John's	442	102
AMAS, THE		Nassau	13,878	351
RBADOS		Bridgetown	430	290
IZE		Belmopan	22,966	388
NADA		Ottawa	9,970,610	35,624
STA RICA		San José	51,100	4,930
BA		Havana	110,861	11,239
MINICA		Roseau	751	74
MINICAN PUBLIC		Santo Domingo	48,511	10,169
SALVADOR		San Salvador	21,041	6,172
ENADA		St George's	344	108
ATEMALA		Guatemala City	108,889	15,461
TI		Port-au-Prince	27,750	10,912
NDURAS		Tegucigalpa	112,088	9,039
MAICA		Kingston	10,991	2,698
XICO		Mexico City	1,958,201	124,575
CARAGUA		Managua	129,494	6,026
NAMA		Panamá	75,517	3,753
KITTS & NEVIS		Basseterre	261	53
LUCIA		Castries	539	180
VINCENT & E GRENADINES		Kingstown	388	110
ITED STATES AMERICA		Washington, DC	9,629,091	326,626
SOUTH AMERICA				
GENTINA		Buenos Aires	2,780,400	44,293
LIVIA		La Paz/Sucre	1,098,581	11,138
AZIL		Brasília	8,514,215	207,353
ILE		Santiago	756,626	17,789
OLOMBIA		Bogotá	1,138,914	47,699
UADOR		Quito	283,561	16,291
ENCH GUIANA		Cayenne	90,000	250
YANA		Georgetown	214,969	747
RAGUAY		Asunción	406,752	6,944
RU		Lima	1,285,216	31,037

Country	Flag	Capital	Area (sq km)	Population (thousands)
SURINAME		Paramaribo	163,265	592
TRINIDAD & TOBAGO		Port of Spain	5,130	1,380
URUGUAY		Montevideo	175,016	3,360
VENEZUELA		Caracas	912,050	31,304
AFRICA				
ALGERIA		Algiers	2,381,741	40,969
ANGOLA		Luanda	1,246,700	29,310
BENIN		Porto-Novo	112,622	11,039
BOTSWANA		Gaborone	581,730	2,215
BURKINA FASO		Ouagadougou	274,200	20,108
BURUNDI		Bujumbura	27,834	11,467
CABO VERDE		Praia	4,033	561
CAMEROON		Yaoundé	475,442	24,995
CENTRAL AFRICAN REPUBLIC		Bangui	622,984	5,625
CHAD		Ndjamena	1,284,000	12,076
COMOROS		Moroni	2,235	808
CONGO		Brazzaville	342,000	4,955
CÔTE D'IVOIRE		Yamoussoukro	322,463	24,185
DEMOCRATIC REPUBLIC OF THE CONGO		Kinshasa	2,344,858	83,301
DJIBOUTI		Djibouti	23,200	865
EGYPT		Cairo	1,001,449	97,041
EQUATORIAL GUINEA		Malabo	28,051	778
ERITREA		Asmara	117,600	5,919
ETHIOPIA		Addis Ababa	1,104,300	105,350
GABON		Libreville	267,668	1,772
GAMBIA, THE		Banjul	11,295	2,051
GHANA		Accra	238,533	27,500
GUINEA		Conakry	245,857	12,414
GUINEA-BISSAU		Bissau	36,125	1,792
KENYA		Nairobi	580,367	47,616
LESOTHO		Maseru	30,355	1,958
LIBERIA		Monrovia	111,369	4,689
LIBYA		Tripoli	1,759,540	6,653
MADAGASCAR		Antananarivo	587,041	25,054

population figures are 2018 estimates where available

Country	Flag	Capital	Area (sq km)	Population (thousands)
MALAWI		Lilongwe	118,484	19,196
MALI		Bamako	1,240,192	17,885
MAURITANIA		Nouakchott	1,025,520	3,759
MAURITIUS		Port Louis	2,040	1,356
MOROCCO		Rabat	446,550	33,987
MOZAMBIQUE		Maputo	801,590	26,574
NAMIBIA		Windhoek	824,292	2,485
NIGER		Niamey	1,267,000	19,245
NIGERIA		Abuja	923,768	190,632
RWANDA		Kigali	26,338	11,901
SÃO TOMÉ & PRÍNCIPE		São Tomé	964	201
SENEGAL		Dakar	196,722	14,669
SEYCHELLES		Victoria	455	94
SIERRA LEONE		Freetown	71,740	6,163
SOMALIA		Mogadishu	637,657	11,031
SOUTH AFRICA		Cape Town/ Pretoria	1,221,037	54,842
SOUTH SUDAN		Juba	620,000	13,026
SUDAN		Khartoum	1,886,086	37,346
SWAZILAND		Mbabane	17,364	1,467
TANZANIA		Dodoma	945,090	53,951
TOGO		Lomé	56,785	7,965
TUNISIA		Tunis	163,610	11,404
UGANDA		Kampala	241,038	39,570
WESTERN SAHARA		El Aaiún	266,000	603
ZAMBIA		Lusaka	752,618	15,972
ZIMBABWE		Harare	390,757	13,805

EUROPE

Country	Flag	Capital	Area (sq km)	Population (thousands)
ALBANIA		Tirana	28,748	3,048
ANDORRA		Andorra La Vella	468	86
AUSTRIA		Vienna	83,859	8,754
BELARUS		Minsk	207,600	9,550
BELGIUM		Brussels	30,528	11,491
BOSNIA-HERZEGOVINA		Sarajevo	51,197	3,856
BULGARIA		Sofia	110,912	7,102

Country	Flag	Capital	Area (sq km)	Population (thousands)
CROATIA		Zagreb	56,538	4
CZECHIA		Prague	78,866	10
DENMARK		Copenhagen	43,094	5
ESTONIA		Tallinn	45,100	1
FINLAND		Helsinki	338,145	5
FRANCE		Paris	551,500	67
GERMANY		Berlin	357,022	80
GREECE		Athens	131,957	10
HUNGARY		Budapest	93,032	9
ICELAND		Reykjavik	103,000	
IRELAND		Dublin	70,273	5
ITALY		Rome	301,318	62
KOSOVO		Pristina	10,887	1
LATVIA		Riga	64,600	1
LIECHTENSTEIN		Vaduz	160	
LITHUANIA		Vilnius	65,200	2
LUXEMBOURG		Luxembourg	2,586	
MACEDONIA		Skopje	25,713	2
MALTA		Valletta	316	
MOLDOVA		Kishinev	33,851	3
MONACO		Monaco	1	
MONTENEGRO		Podgorica	14,026	
NETHERLANDS		The Hague	41,526	17
NORWAY		Oslo	323,877	5
POLAND		Warsaw	323,250	38
PORTUGAL		Lisbon	88,797	10
ROMANIA		Bucharest	238,391	21
RUSSIA		Moscow	17,075,400	142
SAN MARINO		San Marino	61	
SERBIA		Belgrade	77,474	7
SLOVAKIA		Bratislava	49,012	5
SLOVENIA		Ljubljana	20,256	1
SPAIN		Madrid	497,548	48
SWEDEN		Stockholm	449,964	9

Country	Flag	Capital	Area (sq km)	Population (thousands)
TZERLAND		Bern	41,284	8,236
RAINE		Kiev	603,700	44,034
TED KINGDOM		London	241,857	64,769
ICAN CITY		Vatican City	0.44	1

ASIA

Country	Flag	Capital	Area (sq km)	Population (thousands)
GHANISTAN		Kabul	652,090	34,125
MENIA		Yerevan	29,800	3,045
ERBAIJAN		Baku	86,600	9,961
HRAIN		Manama	694	1,411
NGLADESH		Dhaka	143,998	157,827
UTAN		Thimphu	47,000	758
UNEI		Bandar Seri Begawan	5,765	444
MBODIA		Phnom Penh	181,035	16,204
INA		Beijing	9,596,961	1,379,303
PRUS		Nicosia	9,251	1,222
ST TIMOR		Dili	14,874	1,291
ORGIA		Tbilisi	69,700	4,926
DIA		New Delhi	3,287,263	1,281,936
DONESIA		Jakarta	1,904,569	260,581
AN		Tehran	1,648,195	82,022
AQ		Baghdad	438,317	39,192
RAEL		Jerusalem	20,600	8,300
PAN		Tokyo	377,829	126,451
RDAN		Amman	89,342	10,248
ZAKHSTAN		Astana	2,724,900	18,557
WAIT		Kuwait City	17,818	2,875
RGYZSTAN		Bishkek	199,900	5,789
OS		Vientiane	236,800	7,127
BANON		Beirut	10,400	6,230
ALAYSIA		Kuala Lumpur	329,758	31,382
ALDIVES		Malé	298	393
ONGOLIA		Ulan Bator	1,566,500	3,068
YANMAR		Naypyidaw	676,578	55,124
EPAL		Katmandu	147,181	29,384

Country	Flag	Capital	Area (sq km)	Population (thousands)
NORTH KOREA		Pyŏngyang	120,538	25,248
OMAN		Muscat	309,500	3,424
PAKISTAN		Islamabad	796,095	204,925
PHILIPPINES		Manila	300,000	104,256
QATAR		Doha	11,437	2,314
SAUDI ARABIA		Riyadh	2,149,690	28,572
SINGAPORE		Singapore City	683	5,889
SOUTH KOREA		Seoul	99,268	51,181
SRI LANKA		Colombo	65,610	22,409
SYRIA		Damascus	185,180	18,029
TAIWAN		Taipei	35,980	23,508
TAJIKISTAN		Dushanbe	143,100	8,469
THAILAND		Bangkok	513,115	68,414
TURKEY		Ankara	774,815	80,845
TURKMENISTAN		Ashkhabad	488,100	5,351
UNITED ARAB EMIRATES		Abu Dhabi	83,600	5,927
UZBEKISTAN		Tashkent	447,400	29,749
VIETNAM		Hanoi	331,689	96,160
YEMEN		Sana'	527,968	28,037

AUSTRALIA & THE PACIFIC

Country	Flag	Capital	Area (sq km)	Population (thousands)
AUSTRALIA		Canberra	7,741,220	23,232
FIJI		Suva	18,274	921
KIRIBATI		Tarawa	726	108
MARSHALL ISLANDS		Majuro	181	75
MICRONESIA		Palikir	702	104
NAURU		Yaren	21	10
NEW ZEALAND		Wellington	270,534	4,510
PALAU		Melekeok	459	21
PAPUA NEW GUINEA		Port Moresby	462,840	6,910
SAMOA		Apia	2,831	200
SOLOMON ISLANDS		Honiara	28,896	648
TONGA		Nuku'alofa	650	106
TUVALU		Fongafale	30	11
VANUATU		Port-Vila	12,189	283

THE SOLAR SYSTEM

This diagram shows the planets of the Solar System according to sizes and position relative to the Sun

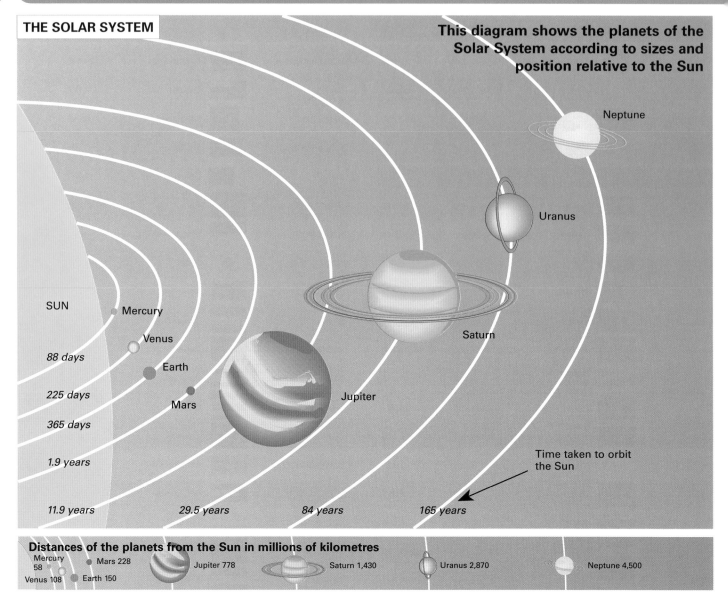

SUN — Mercury

88 days

Venus

225 days

Earth

365 days

Mars — Jupiter

1.9 years

Saturn

Uranus

Neptune

11.9 years   29.5 years   84 years   165 years

Time taken to orbit the Sun

Distances of the planets from the Sun in millions of kilometres

Mercury 58   Mars 228   Jupiter 778   Saturn 1,430   Uranus 2,870   Neptune 4,500
Venus 108   Earth 150

• The Universe is made of many galaxies, or collections of stars. Earth's galaxy is called the Milky Way. It is made up of about 100,000 million stars. The Sun is one of these stars.

• Around the Sun revolve eight planets, one of which is the Earth. The Earth is the fifth largest planet. The Sun, its planets and their satellites are known as the Solar System.

• The Sun is the only source of light and heat in the Solar System. Other planets are visible from the Earth because of the sunlight which they reflect.

• The planets orbit the Sun in the same direction – anti-clockwise when viewed from the northern hemisphere. They also rotate on their own axes.

• The planets remain in orbit because they are attracted by the Sun's pull of gravity. The Earth takes 365 days (a year) to go round the Sun.

THE MOON

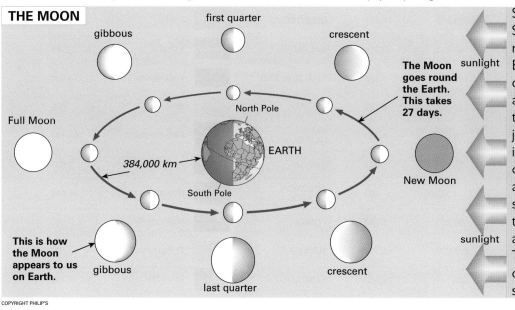

first quarter

gibbous   crescent

The Moon goes round the Earth. This takes 27 days.   sunlight

North Pole

Full Moon

384,000 km   EARTH

South Pole

New Moon

This is how the Moon appears to us on Earth.

gibbous   last quarter   crescent

sunlight

Some planets of the Solar System have satellites that revolve around them. The Earth has just one satellite, called the Moon. The Moon is about a quarter of the size of the Earth. It orbits the Earth in just over 27 days. The Moon is round but we on Earth see only the parts lit by the Sun and we never see 'the dark side'. This makes it look as if the Moon is a different shape at different times of the month. These are known as the phases of the Moon. Phases are shown in this diagram.

• Seasons happen because the Earth's axis is tilted at an angle of 23½°. The Earth revolves around the sun. This gives us the seasons of the year.

• In June, the northern hemisphere is tilted towards the Sun. As a result, it receives more hours of sunshine in a day and therefore has its warmest season, summer.

• Six months later, in December, the Earth has moved halfway round the Sun so that the southern hemisphere is tilted towards the Sun and it has its summer.

THE YEAR & SEASONS

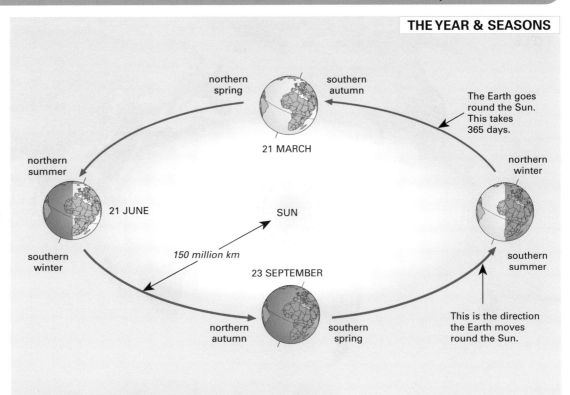

HOW THE LANDSCAPE RESPONDS TO THE SEASONS

In northern latitudes, the seasons each last about three months. Spring arrives in March, Summer in June, Autumn in September, and Winter in December. In the USA, Autumn is known as Fall, as leaves fall from trees in this season.

In southern latitudes (for example, the southern zone of South America, or Australia), the reverse is the case. Spring arrives in September, Summer in December, Autumn in March, and Winter in June.

DAY & NIGHT

• From Earth, the Sun appears to rise in the east, reach its highest point at noon, and then set in the west. In reality, it is not the Sun that is moving but the Earth has rotated from west to east.

• Due to the tilting of the Earth, the length of day and night varies. In June, the Arctic has constant daylight and the Antarctic has constant darkness. The situations are reversed in December. In the Tropics, the length of day and night varies little.

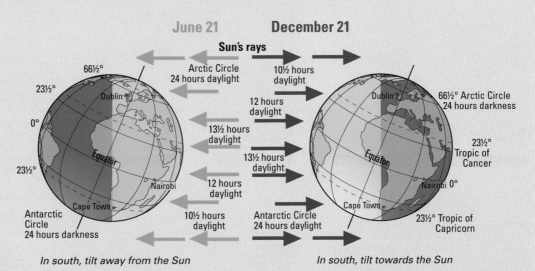

CLIMATE TYPES

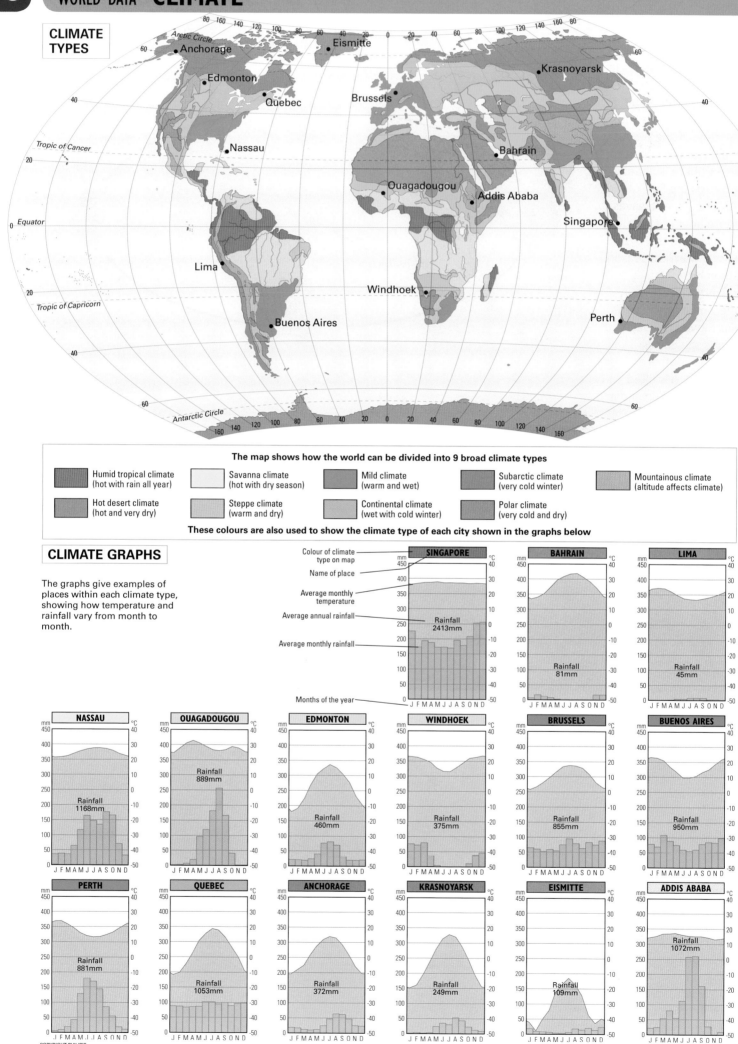

The map shows how the world can be divided into 9 broad climate types

Humid tropical climate (hot with rain all year)

Hot desert climate (hot and very dry)

Savanna climate (hot with dry season)

Steppe climate (warm and dry)

Mild climate (warm and wet)

Continental climate (wet with cold winter)

Subarctic climate (very cold winter)

Polar climate (very cold and dry)

Mountainous climate (altitude affects climate)

These colours are also used to show the climate type of each city shown in the graphs below

CLIMATE GRAPHS

The graphs give examples of places within each climate type, showing how temperature and rainfall vary from month to month.

Colour of climate type on map

Name of place

Average monthly temperature

Average annual rainfall

Average monthly rainfall

Months of the year

SINGAPORE — Rainfall 2413mm

BAHRAIN — Rainfall 81mm

LIMA — Rainfall 45mm

NASSAU — Rainfall 1168mm

OUAGADOUGOU — Rainfall 889mm

EDMONTON — Rainfall 460mm

WINDHOEK — Rainfall 375mm

BRUSSELS — Rainfall 855mm

BUENOS AIRES — Rainfall 950mm

PERTH — Rainfall 881mm

QUEBEC — Rainfall 1053mm

ANCHORAGE — Rainfall 372mm

KRASNOYARSK — Rainfall 249mm

EISMITTE — Rainfall 109mm

ADDIS ABABA — Rainfall 1072mm

COPYRIGHT PHILIP'S

Humid tropical climate

Hot desert climate

Savanna climate

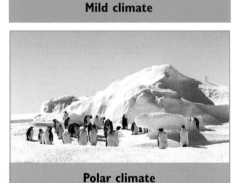

Mild climate

Polar climate

Mountainous climate

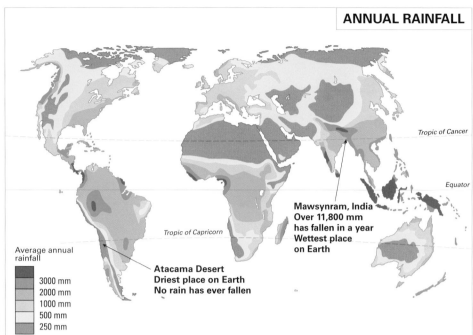

ANNUAL RAINFALL

Tropic of Cancer

Equator

Mawsynram, India
Over 11,800 mm
has fallen in a year
Wettest place
on Earth

Tropic of Capricorn

Average annual
rainfall

- 3000 mm
- 2000 mm
- 1000 mm
- 500 mm
- 250 mm

Atacama Desert
Driest place on Earth
No rain has ever fallen

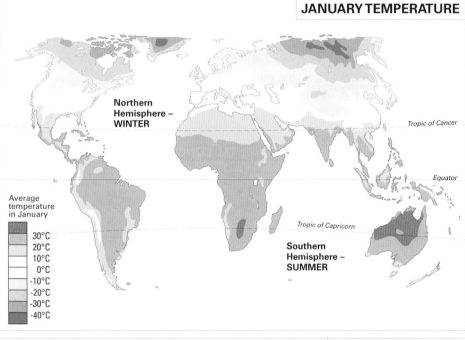

JANUARY TEMPERATURE

Northern
Hemisphere –
WINTER

Tropic of Cancer

Equator

Average
temperature
in January

- 30°C
- 20°C
- 10°C
- 0°C
- -10°C
- -20°C
- -30°C
- -40°C

Tropic of Capricorn

Southern
Hemisphere –
SUMMER

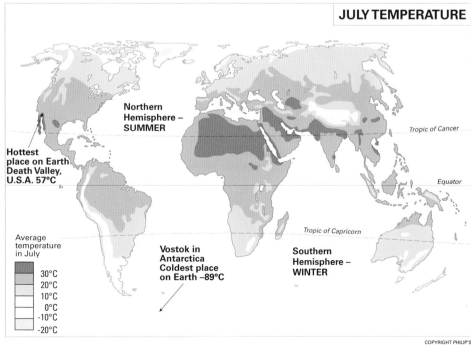

JULY TEMPERATURE

Northern
Hemisphere –
SUMMER

Tropic of Cancer

Hottest
place on Earth
Death Valley,
U.S.A. 57°C

Equator

Average
temperature
in July

- 30°C
- 20°C
- 10°C
- 0°C
- -10°C
- -20°C

Vostok in
Antarctica
Coldest place
on Earth –89°C

Tropic of Capricorn

Southern
Hemisphere –
WINTER

NATURAL VEGETATION

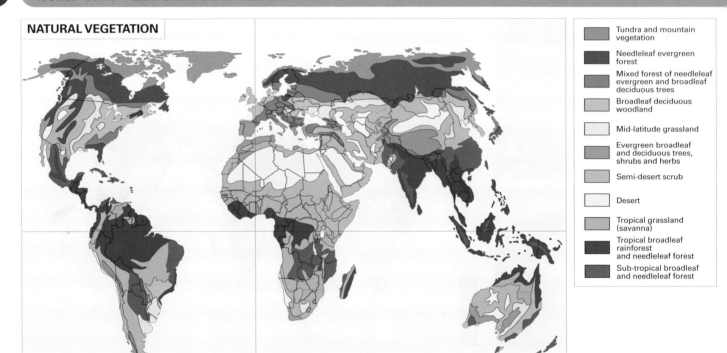

Tundra and mountain vegetation

Needleleaf evergreen forest

Mixed forest of needleleaf evergreen and broadleaf deciduous trees

Broadleaf deciduous woodland

Mid-latitude grassland

Evergreen broadleaf and deciduous trees, shrubs and herbs

Semi-desert scrub

Desert

Tropical grassland (savanna)

Tropical broadleaf rainforest and needleleaf forest

Sub-tropical broadleaf and needleleaf forest

The Natural Vegetation map shows the type of vegetation that would grow if people and animals did not interfere. People have cleared forests and natural grasslands for thousands of years. For example, most of the broadleaf deciduous woodland that would naturally cover parts of Northern Europe cleared for farming and for building.

The Natural Disasters map shows places where populations are affected by drought, floods and insects. It also shows areas where the environment is becoming degraded. The Greenhouse Effect diagram shows how man's activities are changing the world's climate.

NATURAL DISASTERS & ENVIRONMENTAL CONCERNS

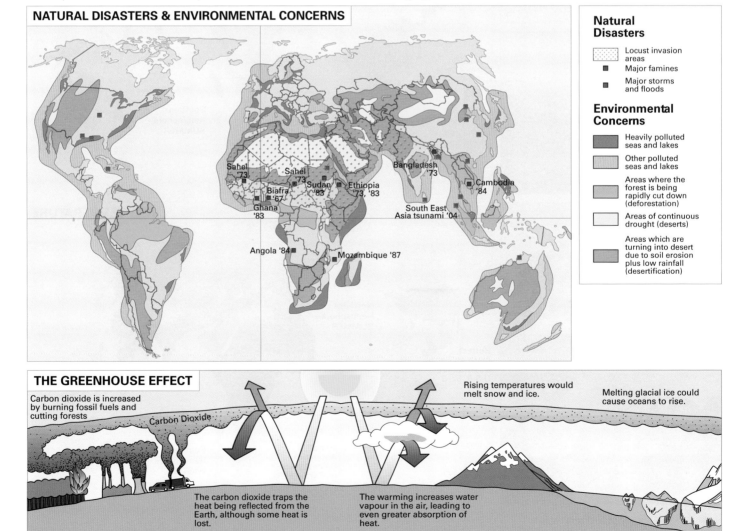

Natural Disasters

Locust invasion areas

Major famines

Major storms and floods

Environmental Concerns

Heavily polluted seas and lakes

Other polluted seas and lakes

Areas where the forest is being rapidly cut down (deforestation)

Areas of continuous drought (deserts)

Areas which are turning into desert due to soil erosion plus low rainfall (desertification)

THE GREENHOUSE EFFECT

Carbon dioxide is increased by burning fossil fuels and cutting forests

Carbon Dioxide

Rising temperatures would melt snow and ice.

Melting glacial ice could cause oceans to rise.

The carbon dioxide traps the heat being reflected from the Earth, although some heat is lost.

The warming increases water vapour in the air, leading to even greater absorption of heat.

PLATE BOUNDARIES & VOLCANOES

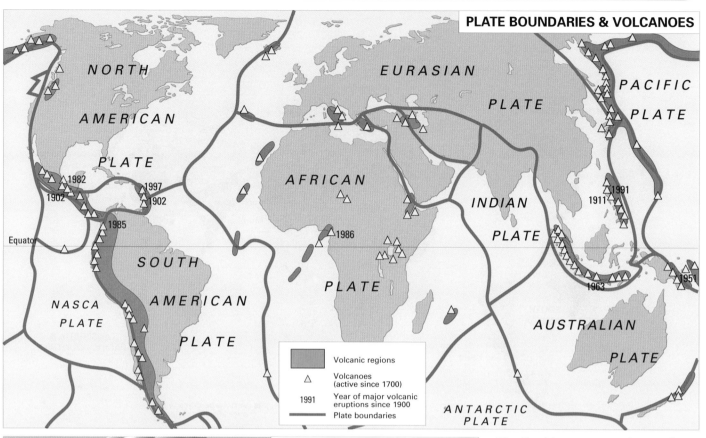

NORTH AMERICAN PLATE

EURASIAN PLATE

PACIFIC PLATE

AFRICAN

INDIAN PLATE

NASCA PLATE

SOUTH AMERICAN PLATE

AUSTRALIAN PLATE

ANTARCTIC PLATE

Equator

	Volcanic regions
△	Volcanoes (active since 1700)
1991	Year of major volcanic eruptions since 1900
——	Plate boundaries

CROSS SECTION THROUGH VOLCANO

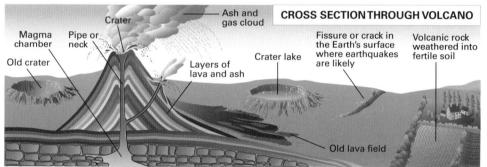

Crater — Ash and gas cloud

Magma chamber
Pipe or neck
Old crater
Layers of lava and ash
Crater lake
Fissure or crack in the Earth's surface where earthquakes are likely
Volcanic rock weathered into fertile soil
Old lava field

The Earth's crust is made up of plates as shown on the map above. Movement of these plates against each other results in earthquakes. The edge of plates can be forced down to great depths and form fluid magma. Volcanoes erupt when this magma is forced upwards to the surface of the Earth.

ZONES OF EARTHQUAKE RISK

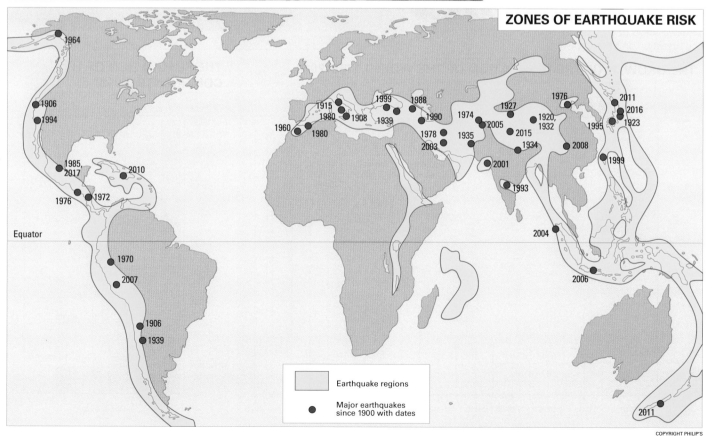

Equator

| | Earthquake regions |
| ● | Major earthquakes since 1900 with dates |

WHERE PEOPLE LIVE

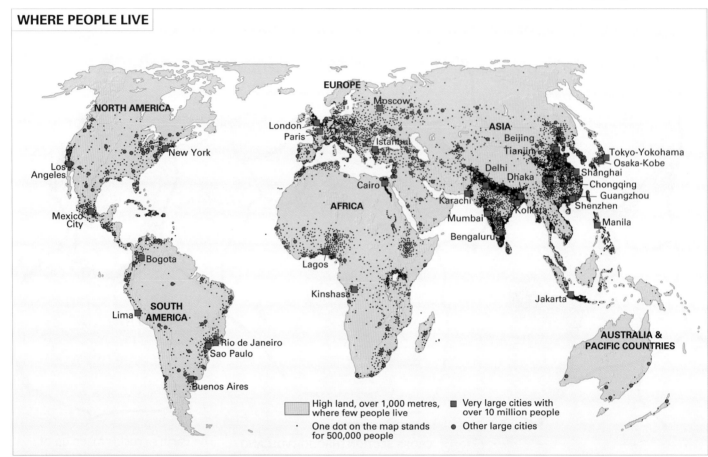

High land, over 1,000 metres, where few people live

■ Very large cities with over 10 million people

· One dot on the map stands for 500,000 people

● Other large cities

Greater Tokyo (the Tokyo-Yokohama metropolitan area) is the largest urban area on Earth.

THE WORLD'S LARGEST CITIES

CITY	CONTINENT	COUNTRY	POPULATION
Tokyo-Yokohama	Asia	Japan	39,800,000
Delhi	Asia	India	27,200,000
Shanghai	Asia	China	24,500,000
Manila	Asia	Philippines	24,100,000
Mumbai	Asia	India	23,600,000
São Paulo	South America	Brazil	21,900,000
Mexico City	North America	Mexico	21,200,000
Beijing	Asia	China	21,200,000

THE GROWTH OF THE POPULATION OF THE WORLD 1750–2018

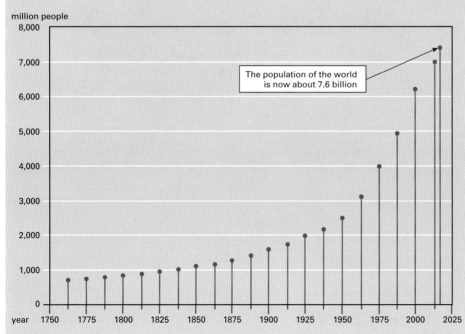

million people

The population of the world is now about 7.6 billion

THE POPULATION OF THE CONTINENTS (2018)

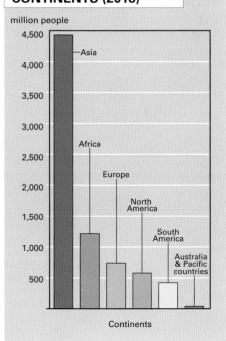

million people

Continents

FISHING & LAND USE

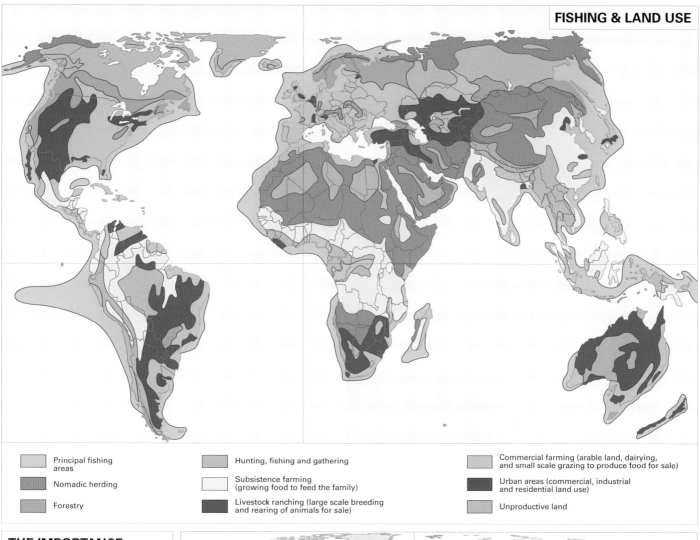

Principal fishing areas	Hunting, fishing and gathering	Commercial farming (arable land, dairying, and small scale grazing to produce food for sale)
Nomadic herding	Subsistence farming (growing food to feed the family)	Urban areas (commercial, industrial and residential land use)
Forestry	Livestock ranching (large scale breeding and rearing of animals for sale)	Unproductive land

THE IMPORTANCE OF FARMING

Percentage of the people who work in farming (2017)

- Over half of the people work in farming
- Between a quarter and a half of the people work in farming
- Between one in ten and a quarter of the people work in farming
- Less than one in ten of the people work in farming
- No data

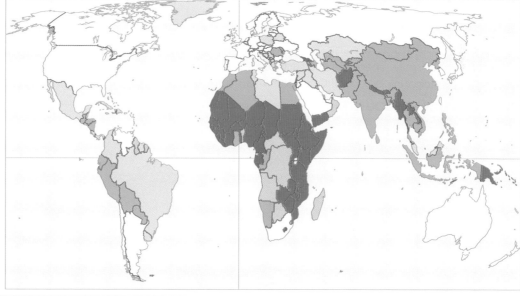

A hundred years ago about 80% of the world's population worked in farming. Today it is only about 30% but farming is still very important in some countries.

METHODS OF FISHING AT SEA

Methods of fishing

There are two types of sea fishing:

1. Deep-sea fishing using large trawlers which often stay at sea for many weeks.

2. Inshore fishing using small boats, traps and nets up to 70 km from the coast.

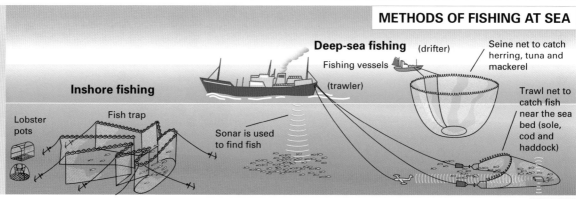

Deep-sea fishing (drifter)

Fishing vessels

(trawler)

Seine net to catch herring, tuna and mackerel

Inshore fishing

Lobster pots

Fish trap

Sonar is used to find fish

Trawl net to catch fish near the sea bed (sole, cod and haddock)

COPYRIGHT PHILIP'S

THE UNITED NATIONS

The United Nations organisation (UN) was established in 1945 to promote worldwide peace and co-operation. Its membership is 193 independent countries and its annual budget is around US$5 billion. Each member of the General Assembly has one vote. The Secretariat is the UN headquarters' administrative arm. The UN has 16 sectoral agencies, headquartered in the US, Canada, France, Switzerland, and other countries, which help members in sectors such as education (UNESCO), agriculture (FAO), health (WHO) and finance (IFC). The UN also has special offices such as those focussed on refugees (UNHCR) or on children (UNICEF).
• Can you find the purpose of all the agencies whose acronyms appear in this chart?

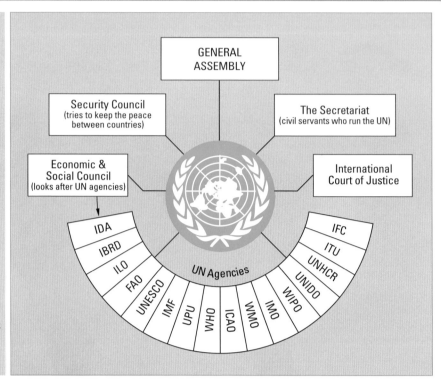

THE COMMONWEALTH

• The Commonwealth of Nations is an association of independent countries, most of which were once colonised by the British. Its objective is to strengthen democratic processes and to support development and economic growth.
• Several Commonwealth countries recognise the British monarch as their head of state. Most use English as one of their languages. There were 53 members in 2018, eleven of these in the Caribbean.
• UK Overseas Territories are also considered as part of the Commonwealth family. There are five such territories in the Caribbean, participating in several Commonwealth activities.

COMMONWEALTH COUNTRIES: CARIBBEAN REGION

COMMONWEALTH COUNTRIES

Commonwealth member

COPYRIGHT PHILIP'S

GROWTH OF THE EU

€ Euro-zone ○ EU headquarters

▮	Founder members (Treaty of Rome 1957)
▮	Joined in 1973
▮	Joined in 1981
▮	Joined in 1986
▮	Joined in 1990 (German unification)
▮	Joined in 1995
▮	Joined in 2004
▮	Joined in 2007
▮	Joined in 2013

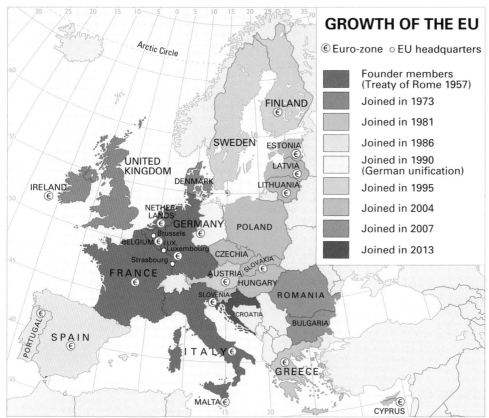

EUROPEAN UNION (EU)

• The European Union evolved from the European Community in 1993. Together, the 28 members states aim to integrate economies, coordinate social developments, and bring about political union.
• 19 of the 28 EU countries have created a currency union referred to as the Euro zone. The countries agreed to end use of their own currencies and adopt the European currency, the Euro, for all transactions. Euro zone members are shown on the map on the left.

ACP STATES

AFRICAN, CARIBBEAN & PACIFIC GROUP OF STATES (ACP)

Formed in 1963, the 79 member states benefit from economic ties with the EU.

ACP states in the Caribbean

Antigua & Barbuda	Guyana
Bahamas	Haiti
Barbados	Jamaica
Belize	St. Kitts & Nevis
Cuba	St. Lucia
Dominica	St. Vincent & the Grenadines
Dominican Republic	Suriname
Grenada	Trinidad & Tobago

• All countries of the world committed themselves in 2015 to *Transforming our World: the 2030 Agenda for Sustainable Development.*
• The diagram shows the 17 higher-order goals which cover a broad range of social and economic development issues.
• These include poverty, hunger, health, education, climate change, gender equality, water, sanitation, energy, environment and social justice.
• The goals are known as Sustainable Development Goals or SDGs for short.

THE WORLD'S SUSTAINABLE DEVELOPMENT GOALS

• Which of the goals are important for your country?

• Which one would you select as the most important of all?

How the index is organised

The index enables you to quickly find places in the atlas. You may be looking for a physical feature, such as an island, a mountain peak, a river or a lake. Or you may be looking for a man-made feature such as a country, a city or a town.

All names in the index are listed in alphabetical order. Some names include a descriptor for the kind of feature it is (for example, the Gulf of Paria, to the west of Trinidad), the name is in alphabetical order followed by the description:

Paria, Gulf of

Sometimes, the same name occurs in more than one country. In these cases, the country names are added after each place name, and they are in the index alphabetically by country. For example:

Richmond, *Jamaica*
Richmond, *U.S.A.*

All river names are shown in blue (without the word river). For example, the Demerara river in Guyana:

Demerara

Every name in the index is followed by the page number of the map it appears on, a letter and then a number. For example:

Chaguanas **48** C2

The best map to find Chaguanas is on page 48. C2 is its grid reference. This is explained on the opposite page.

A lower case letter after the page number refers to the small map on that page.

Using grid references to find places

The main map for each region or country is divided into grid rectangles by the lines of latitude and longitude. For these rectangles, columns are labelled with letters and rows with numbers, found in small yellow boxes at the sides of a map. The index suggests that Chaguanas can be found at 48 C2. This means that on page 48, Chaguanas is somewhere in the grid rectangle where column C crosses row 2.

A place is always indexed to the grid rectangle where the town symbol is, rather than where the name is. Rivers may cross many grid rectangles, so the index has the rectangle where the name of the river nearest the sea is printed.

An area such as an island, mountain range or country may fall into more than one grid rectangle. The rectangle named in the index is the centre of the area. For example, the parish of Hanover in Jamaica is in the index as Hanover 26 B2.

To find Hanover, open your atlas on page 26, place the index finger of your left hand on 2 and the index finger of your right hand on B. Move your fingers across and up until they meet. You will have found Hanover.

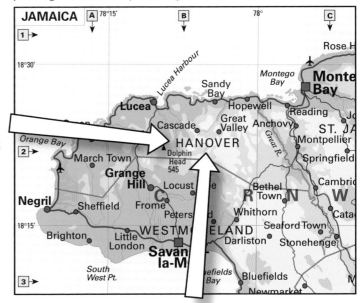